This may give you an insight into some of my weird ideas!

July 1993

SCIENCE:
The Very Idea

D0082690

KEY IDEAS

Series Editor: PETER HAMILTON
The Open University, Milton Keynes

Designed to complement the successful *Key Sociologists*, this series covers the main concepts, issues, debates, and controversies in sociology and the social sciences. The series aims to provide authoritative essays on central topics of social science, such as community, power, work, sexuality, inequality, beliefs and ideology, class, family, etc. Books adopt a strong individual 'line' constituting original essays rather than literary surveys, and form lively and original treatments of their subject matter. The books will be useful to students and teachers of sociology, political science, economics, psychology, philosophy, and geography.

THE SYMBOLIC CONSTRUCTION OF COMMUNITY
ANTHONY P. COHEN, Department of Social Anthropology, University of Manchester

SOCIETY
DAVID FRISBY and DEREK SAYER, Department of Sociology, University of Glasgow

SEXUALITY
JEFFREY WEEKS, Social Work Studies Department, University of Southampton

WORKING
GRAEME SALAMAN, Faculty of Social Sciences, The Open University, Milton Keynes

BELIEFS AND IDEOLOGY
KENNETH THOMPSON, Faculty of Social Sciences, The Open University, Milton Keynes

EQUALITY
BRYAN TURNER, School of Social Sciences, The Flinders University of South Australia

HEGEMONY
ROBERT BOCOCK, Faculty of Social Sciences, The Open University, Milton Keynes

SCIENCE: The Very Idea
STEVE WOOLGAR, Department of Sociology, Brunel University, Uxbridge

SCIENCE:
The Very Idea

STEVE WOOLGAR
Department of Sociology
Brunel University

ELLIS HORWOOD LIMITED
Publishers · Chichester

TAVISTOCK PUBLICATIONS
London and New York

First published in 1988 by
ELLIS HORWOOD LIMITED
Market Cross House, Cooper Street,
Chichester, Sussex, PO19 1EB, England
and

TAVISTOCK PUBLICATIONS LIMITED
11 New Fetter Lane, London EC4P 4EE

Published in the USA by
TAVISTOCK PUBLICATIONS
and ELLIS HORWOOD LIMITED
in association with METHUEN INC.
29 West 35th Street, New York, NY 10001–2291

British Library Cataloguing in Publication Data
Woolgar, Steve 1950–
Science: the very idea. — (Key ideas).
1. Science — sociological perspectives
I. Title   II. Series
306'.45

Library of Congress CIP data available

ISBN 0–7458–0041–6 (Ellis Horwood Limited — Library Edn.)
ISBN 0–7458–0042–4 (Ellis Horwood Limited — Student Edn.)

Phototypeset in Times by Ellis Horwood Limited
Printed and bound in Great Britain by Richard Clay Ltd., Chichester, West Sussex

Contents

Steve Woolgar is Lecturer in Sociology at Brunel University, a position he has held since 1975. From 1983–84 he was a Research Fellow at the Massachusetts Institute of Technology, USA; and was Visiting Professor at McGill University, Canada, from 1979–81. He was awarded a Ph.D. (1978) and an M.A. (1976) on the Sociology of Science, and a B.A. (1972) on Engineering Tripos, all from Emmanuel College, University of Cambridge. Dr Woolgar has written two previous books.

Editor's foreword

Sociological studies of knowledge and science have always laboured under the burden of a widespread assumption that science represents pure and 'objective' knowledge — ideally untainted by 'social' factors which distort or divert it from its real purpose, the uncovering of the 'true' nature of the physical world.

As a result the term science itself has come to denote a privileged realm of knowledge, removed and seemingly sharply differentiated from other intellectual domains. This is a process which has been in train since at least the thirteenth century in Western civilization, and it has led to a point where science and technology are represented in much of popular and high culture as playing a hegemonic role as the moving forces of social and economic change. Indeed, to describe a particular proposition, concept or theory as 'unscientific', is effectively to condemn it.

This has meant that many intellectual activities whose objects and methods may diverge from those of the natural sciences are nonetheless required to conform to certain 'scientific' standards. Sociology itself has struggled with the question of what it is to be a social science, and the work of many of its theorists has been marked by the encounter.

Steve Woolgar's study is the product of a new approach to science as a social activity. Rather than treating the social context of science as something divorced or at least separable from its practice, Woolgar's work represents a view of science which finds little to demarcate it from other social activities save the efforts of its practitioners to create and service such a demarcation.

As will become clear this position is one which leads to a certain degree of scepticism about the knowledge claims of sciences. However, it is also helpful — precisely because it serves to answer unresolved questions about the scientific status of sociology and to indicate that the problems of objectivity and subjectivity to which sociology is supposedly exposed are ultimately partial rather than general.

Dr Woolgar writes from a perspective which has come to be known as the 'social study of science' so as to differentiate its concerns from other and older traditions such as sociology of science (where writers such as Robert K. Merton† have developed a detailed analysis of science focusing on its interdependence with other social institutions), sociology of knowledge where concerns with ideology and the sociocultural bases of sciences have been pre-eminent, and the more 'social' interests of historians and philosophers of science, like Kuhn and Feyerabend.

Lucidly argued, authoritative and instructive about the institution and practice of science, Dr Woolgar's book deserves to be read by scientists of any persuasion — social or natural — and by anybody with the slightest interest in how science and scientific thinking operate.

The social study of science is set to trigger a radical change in our understanding of science — perhaps a post-Kuhnian revolution, to borrow a concept from an earlier and equally influential shift in the way science and its practice are understood.

Peter Hamilton

†See the Key Sociologists book, *Robert K. Merton,* by Charles Crothers (Ellis Horwood, Chichester; Tavistock, London, 1987), which discusses his contribution to sociology of science.

Preface

One of the most remarkable features of modern thought is the extent to which ideas about science have changed. In the last twenty years, a number of different disciplines have challenged traditional views about science. Although critical questions about the politics and impact of science have a longer pedigree, it is only comparatively recently that critical attention has been directed towards the 'internal' workings of science. Longstanding views of the production of scientific knowledge, the textbook (or 'fairy story') image of science, have been called into question. Now the practice of science is itself the object of critical scrutiny.

Challenges to the idea of science as a privileged way of producing reliable knowledge can be found in different forms in several disciplines: literary theory, philosophy, history, anthropology and sociology. In addition, several significant intellectual movements cut across these traditionally defined boundaries and provide the basis for an interdisciplinary critique: deconstruction, the critique of representation, structuralism and post-structuralism, relativism, post-modernism. Not surprisingly, criticisms of science evolved by these developments have radical and far-reaching consequences. For example, the critique of the production of scientific knowledge has implications for discussions about the impact of science and its politics. More importantly, the same critique affects fundamental preconceptions about science, especially as these underpin most bodies of modern scholarship.

This book provides an introduction to recent challenges to the idea of science and aims to sketch out some of the more radical implications of this approach. It concentrates in particular on the claims and achievements of the social study of science, a recent body of work that has been influential in transforming traditionally held attitudes towards science. This perspective has been as vigorous in its production of research as it is contentious in its epistemological preconceptions. But while influential amongst historians and (to a lesser extent) philosophers of science, the more general implications of this perspective for scholarship as a whole are yet to be recognized. Thus, for example, many introductions to social science disciplines continue to engage in tired versions of whether or not social science is 'really' a science, apparently oblivious to the insights of recent work.

The book aims to make recent work accessible to those whose special interests are in areas other than science. The book provides a critical overview of our understanding of the institution and practice of science; it outlines the achievements and deficiencies of recent research in the social study of science; it indicates some of the important implications of this work for a range of areas of scholarship in the social sciences which take their cue (either implicitly or explicitly) from prevailing conceptions of science and the scientific method.

Although the Notes at the end of each chapter, and the Suggestions for Further Reading contain a fair proportion of the main sources which could be drawn upon in introducing the topic, my aim is to provide more than just a textbook survey. My strategy is to review this work from a particular argumentative position. One main line of argument is that despite prolific research activity in the social study of science, undertaken in a general climate of increasing scepticism about the claims and achievements of the natural sciences, the social sciences have yet fully to come to terms with the very idea of science. In particular I try to outline the deleterious consequences of advancing partial critiques of science which themselves continue to trade upon unexplicated notions of scientific practice in their own work. I want to examine some of the reasons for this state of affairs and to consider briefly the prospects for a further radical shift (a post-Kuhnian revolution?) in our understanding of the phenomenon of science.

In short, this volume is intended as an essayistic general introduction to the notion of science which argues a firm line with respect to current ideas. It is designed for sociologists and social scientists, both teachers and researchers, who are specialists in areas other than the sociology of science and who will benefit from an introductory demonstration of both the significance and dangerous potential of this work for social science as a whole.

My main debt is to my many colleagues in the social study of science, whose work has been a constant source of stimulation. Unhappily, the form of argument here requires the selection from a substantial body of scholarship, so that my indebtedness is only partially reflected in my use of particular examples and citations. Hopefully, my efforts to introduce the arguments will encourage others to delve deeper and to decide for themselves the adequacy of my re-presentations. I would also like to thank Peter Hamilton for his encouragement and considerable forbearance.

Introduction

The notion of science has been a dominant and enduring feature of Western thought since at least the time of Bacon. It could be said that in one way or another, 'science' permeates every aspect of modern life. This is evident in terms of the large investment in science by Western (and, increasingly, by Eastern and third world) societies, in debates over the impact of science on society and so on. These are reflections of the value attached to science as a means of producing reliable knowledge. As a consequence of its prestige and value, science is well developed, highly funded and protected: a major and powerful institution in modern societies.

The high esteem associated with the idea of science is also pervasive in senses less obvious than the proportion of Gross National Product set aside for its pursuit by highly trained professionals in industry and academia. It is a measure of the dominance of ideas about science that there are intense squabbles within and between different intellectual traditions about the scientific standing of their own knowledge practices. And at a more general level, we can identify the influence of ideas about science in our mundane considerations of knowing, being certain or sure, getting proof, obtaining evidence, winning an argument and so on. Adverts try to persuade us of the attractions of a new brand of toothpaste based upon scientific evidence. We are told we are less likely to die of ignorance if we base our actions upon reliable knowledge. Moreover, our participation in everyday life obliges us routinely to engage in the practice which lies at the very heart of science: representation. Leaving aside for the moment the question of the relative merits of representational practice in science and non-science, there is an obvious sense in which our mundane practices mimic this feature of science. We describe, account, report. When writing, we reckon to employ a means of representation which enables us to concretize our experiences and observations. Or when speaking, we attempt to communicate knowledge of events and objects, including those outside our immediate experience.

The pervasiveness of ideas about science, their influence in almost all aspects of modern existence, has two major consequences for any project which attempts to study the phenomenon of science. *Firstly,* the social study of science (SSS) has a special strategic significance in that it attempts to account for science and technology at a number of different levels [1]. It looks at both the institution and practice of science; not just at social relationships between practitioners, networks of communication, the reward system, the influence of patronage and sponsors, but also at what actually goes on in science on a day-to-day basis. What do scientists do when they are in the laboratory? It is concerned not just with the social arrangements and organization of science as a social institution, but also with science as a cultural phenomenon. In other words, the social study of science attempts an understanding of a belief system which extends far beyond the formal social organization of science, and far outside the walls of the laboratory. What accounts for the extraordinary reliance upon, and widespread support for, science as a credible and highly valued means of producing knowledge? If the physicist at his laboratory bench is only the tip of the iceberg, a tangible manifestation of a pervasive belief system, SSS has to be prepared to look further than the interviewee-respondent, to address some of the fundamentals of the belief system itself.

Secondly, the pervasive influence of beliefs and views about science is such that any study, especially if conducted within the context of professional academia, is itself likely to be affected by pretensions to science. The very omnipresence of the thing to be studied will affect how we study it. SSS thus stands in a highly unusual position with respect to the object of its study. For there is at least the potential that our object rebounds upon us. In 'safer' areas of study, the sociology of the family, of education, or deviance, this potential feedback between object and study is entirely absent (or, at least, less immediately apparent). But SSS comes face to face with a series of tricky questions about the most appropriate way to study science. Should we be scientific in our study of science? Can we avoid being scientific when we study science? Do we have to presume something about our object of study even to carry out the study?

These questions become yet more pressing when we consider the main conclusions of SSS: that there is no essential difference between science and other forms of knowledge production; that there is nothing intrinsically special about 'the scientific method'; indeed, even if there is such a thing as the 'scientific method', much scientific practice proceeds in spite of the canons of scientific method rather than because of them. At least one philosopher (Feyerabend) argues that we are unlikely to produce reliable knowledge *unless* we deliberately flout the rules of science. Clearly, the grounds for these conclusions need considerable elaboration before we take them on board. It is nonetheless important to present them in outline form at this early stage to see the consequences for studying science from a 'social' perspective. To the extent that our own standpoint is shot through with presumptions about what it is to be scientific, we cannot continue to separate science from our own efforts at understanding, from what (for us) counts as an adequate account, from the grounds on which our own deliberations might be considered persuasive.

We begin to see, even at this early stage of the argument, that while the social study of science has immense potential and significance, it is a project which is neither safe nor very comfortable. It raises fundamental questions which persistently rebound upon the questioner and prompt awkward doubts about the cool, clinical relationship sometimes imagined to exist between social scientist and the object of study. The analyst can not be detached from his object in any straightforward sense, precisely because it is this distinction, the separation between analysis and object, which is the lynchpin of — the fundamental rationale for — science itself. In its most virulent form, the challenge to the idea of science takes on just this cartesian dualism between self and other, between subject and object, representation and reality.

The radical approach to SSS means we must take issue with the concept of 'science'. But a yet more intriguing consequence of this line of argument is that standard conceptions of 'society' and 'technology' also come under fire. For if the view of science and society as separate analytic objects is to be abandoned, then these conceptions collapse into one another. Whereas in the traditional view, the separation between these three entities is upheld in the study of 'science, technology and society' — and in the corresponding Venn diagram we portray the three domains as interconnected but not overlapping — the radical view maps these entities one on top of the other — and the Venn diagram comprises a single domain. It is this kind of scheme which is behind (and the radical consequence of) calls for abandoning the distinction between the social and the technical, the social and the cognitive and so on. It is not that science has its 'social aspects', thus implying that a residual (hard core) kernel of science proceeds untainted by extraneous non-scientific (i.e. 'social') factors, but that science is itself constitutively social. This means that just as we should abandon the idea of science as a privileged or even just separate domain of activity and inquiry, similarily the notion of 'social' (similarily 'cultural', 'psychological' and so on) has to be substantially modified. The potential consequence of our radical study of science is no less than to make redundant the very concepts of 'social' and 'society'.

To reach this conclusion, the argument is organized along the following lines. Chapter 1 surveys the attempts of philosophy, history and sociology to provide an answer to the familiar question about demarcation: what is science? It is suggested that the extraordinary variation in responses to this question should lead us to treat the idea of 'science' as an evaluative resource, rather than as a definitive entity. Resistance to this suggestion is explained in Chapter 2 in terms of an enduring commitment on the part of the social sciences to a particular conception of representation. Two key policies — inversion and feedbacking — are proposed as a means of countering and/or challenging this commitment. Chapter 3 examines recent attempts to develop a critique of representation by developing a sociology of logic and reason. In particular, we consider the advantages and disadvantages of the 'strong programme in the sociology of scientific knowledge' as a counterweight to the views of objectivist philosophy. Our first application of feedbacking reveals the extent to which the 'strong programme', especially in its programmatic mood, has failed to exorcise its own reliance on rules and logic. Chapter 4 looks more closely at the notion of connection between representation and object. Through the examples of discovery and the construction of facts, it is argued that the standard direction of connection must be inverted: the objects of the natural world can thereby be seen to be accomplished in virtue of scientists' representational practices. This leads us to a consideration, in Chapter 5, of the nature of scientific discourse and, in particular, of the way in which key ideas about representation inform the practices of argument and explanation. The argument here is that the discourse of science is to be understood as a discourse which structures and sustains a particular moral order of relationships between agents of representation, technologies of representation and their represented 'objects'. Chapter 6 then addresses in more detail the problems and possibilities of subverting this moral order. The concept of reflexivity — one possible means of 'keeping alive' the insights of inversion and feedbacking — is introduced by way of a review of attempts to develop 'anthropological' perspectives on scientific practice. Finally, in Chapter 7, the implications of the preceding discussion are brought to bear on the issue of demarcation with which we began. What are the prospects for moving beyond the ideology of representation which lies at the heart of science? And what, in turn, are the implications for social science of the attempts to develop a radical critique of the very idea of science?

NOTE

[1] Terminological note: The term 'social study of science' (SSS) is used here to encompass a range of disciplinary interests in science: notably sociology and history of science, less prominently, philosophy, anthropology and psychology. As we shall see in due course (and, in particular, from Chapter 3 onwards), the sociology of scientific knowledge (SSK) is the term used to denote an especially influential body of scholarship informed by historical and cultural relativism. The relatively narrow disciplinary base of SSK belies the extent of its impact throughout SSS.

1

What is Science?

The social study of science begins with the recognition that science is a highly variable animal. As is well known, there are numerous versions of what science is, of what counts as being scientific. Again, it is testimony to the widespread influence and penetration of ideas about science that competing claims as to what science comprises can be found in many different areas of life. For example, we thus find acrimonious debates about the scientific status of creationism; proclamations by the Secretary of State for Education that the premier funding body for social science research is no longer worthy of the epithet: 'science' [1]; and so on. We begin by distinguishing two major responses - philosophical and historical - to the question of what counts as science.

WHAT IS SCIENCE? - (1) THE PHILOSOPHICAL RESPONSE

In its philosophical treatment, the question about the nature of science has most famously been concerned with *demarcation*. What is it about science that makes it a more reliable knowledge producing system than others? Whereas, for example, belief in God is hard to substantiate except by faith and trust, what about science enables us to place confidence in claims to, say, knowledge about the solar system. In short, what distinguishes science from other activities? Quite obviously, the answer to this question bears directly on attempts by social and human sciences to answer the related question: to what extent should the study of human behaviour emulate the perspectives of the natural sciences. Whether or not the perspective of science is appropriate for the study of social (human, behavioural) phenomena depends on there being something distinctive called 'science' in the first place.

The philosophical quest for demarcation criteria has something of a chequered history [2]. There have been arguments that science is distinctive by virtue of its *results*, but more lately that its *methodology* makes it distinctive. But even agreement on this score conceals considerable differences about what exactly the methodology of science comprises. In the 1950s the principle of *verification* was proposed as the aspect of methodology which distinguished science from non-science. If a statement was capable of being verified, it was said, this was what made it scientific, by virtue of excluding conjectures about beliefs, opinions and preferences and so on. 'God exists', 'I prefer coffee to tea' and so on were categorized as non-scientific statements. But this principle ran into problems on several scores, most notably the logical problem of induction: although tests of verification could be applied to any candidate generalization, the status of such a generalization was always uncertain, in that any next observation may counter it. In other words, verification guaranteed little since, in principle, the generalization could always be undone by the very next observation. Popper's solution was to propose the principle of *falsification* [3]. While no generalization could in principle achieve the status of certainty, Popper suggested that tests of falsification could distinguish the relative merits of generalizations. The essence of scientific methodology, said Popper, is to produce generalizations which resist attempts to falsify them. Attempts should be made to verify instances which counter a proposed generalization; the failure to verify the counter-instance (that is, the failure to falsify), gives (albeit temporary) credence to the generalization. The ever-present uncertainty arising from the logical problem associated with verification, is thus replaced by the promise of an increasingly reliable (although never ultimately certain) generalization which resists more and more attempts to falsify it.

In social sciences, a clear example of falsification occurs in the application of statistical hypothesis testing. Researchers formulate a null hypothesis (for example, there is no statistical association between social class and educational achievement) which they then attempt to falsify. The falsification of the null hypothesis then provides the firmest possible indication (but not proof) that there is indeed a statistical association between the two variables.

Although Popper's proposed solution was a dramatic and much celebrated attempt to break an enduring problem about the status of scientific methodology, it became clear that both verification and falsification suffer from the weakness of the central assumption that observations are neutral: little attention is given to the way in which observations bear upon the statement (generalization) at issue. In the classic example of the generalization 'all swans are white', it may be logically more attractive to attempt its falsification rather than its verification, but both approaches pay scant regard to the issue of what is to count as 'a swan' or 'white'. Is a swan dipped in soot to count as a white swan really? It is a central contention of those more attuned to the communal (social) character of science that decisions about the status of observations, and thence about the applicability of principles like verification or falsification, are made in social context. The whiteness of the swan arises as a result of localized perceptions of whiteness. It is emphatically not an inherent, objective and unambiguous attribute.

Some concessions to the community-based character of observation appear in Lakatos's proposal for a *methodology of scientific research programmes* [4]. By focusing at the level of 'research programme' Lakatos stresses that generalizations (hypotheses, statements) are never assessed in isolation. A research programme comprises a cluster of hypotheses and a series of methodological rules which specify which paths of development to follow and which to avoid. The hypotheses are themselves classified as either belonging to the 'hard core' or the 'protective belt'. Modifications to the hypotheses in the protective belt can make the research programme either 'progressive' or 'degenerating'.

But while the Lakatosian account of science makes Popper's ideas more responsive to the weight of judgement and its effects on the overall development of scientific theories, the methodological rules remain unclear. In particular, it is not apparent that the *rules themselves* can distinguish between progression and degeneration. Indeed it is not clear from historical evidence in what sense such rules exist *at all*, as if ready for consultation by the indecisive scientist. The problematic assumption about the neutrality of observations in verifying or falsifying, is now replaced by a problematic assumption about the determinative characteristic of rules.

This brief survey of philosophical ideas about science highlights the range of variation in philosophers' attempts to specify demarcation criteria for science. The preference of SSS, by contrast with this philosophical quest, is to accept that science can not be distinguished from non-science by decision rules. Judgements about whether or not hypotheses have been verified (or falsified), as to what constitutes the core or periphery in a research programme, and at what point to abandon a research programme altogether, are the upshot of complex social processes within a particular environment. Scientific knowledge does not arise from the application of pre-existing decision rules to particular hypotheses or generalizations.

As we shall later see in more detail, SSS favours the conception of rules as a *post hoc* rationalization of scientific practice rather than as a set of procedures which determine scientific action. This preference supports the contention, highlighted in particular by Kuhn [5], that certain forms of history of science can be considerably misleading. With the benefit of hindsight, historical episodes in science are rewritten so as to conform with certain imputed decision rules. It is assumed in retrospect that these rules *must have* operated to produce the scientific knowledge in question. This retrospective conjuring of the operation of decision rules is done in the light of the current state of knowledge and judgements about the validity or otherwise of the historical claim. It is this 'temporal-centrism' which excludes the messiness of scientific practice, de-emphasizes the uncertainty faced by scientists, omits the red herrings and false trails, and ultimately conveys the impression that the present state of knowledge is the inevitable and logical outcome of historical progression.

A less sceptical view of rules regards them not as out-and-out *post hoc* rationalizations, but as merely one of a number of determinants of scientific action. This line of argument holds that scientific knowledge is *underdetermined* both by observational evidence (the observations themselves do not determine the fate of a statement or generalization) and by decision rules (the prescriptive procedures will not by themselves settle the outcome of a hypothesis). For example, the 'fact' of falsification will not by itself guarantee the rejection of a hypothesis. Underdetermination in this sense clearly allows space for the concomitant influence of additional 'social' factors in the assessment of knowledge claims.

These difficulties with the efforts of philosophers to specify demarcation criteria are supported by the results of recent SSS research into the dynamics of marginal or pseudo-sciences [6]. The central argument of studies of phrenology [7] or of parapsychology [8] is that such work operates in ways quite consistent with the requirements of demarcation. For example, phrenologists can be seen as having acted in a way quite consistent with the opposing view. Or again, only in retrospect is it possible to say that N-rays failed to live up to criteria of falsification.

WHAT IS SCIENCE? - (2) THE HISTORICAL RESPONSE

We see that philosophical attempts to characterize science have generated a variety of demarcation criteria, each of which is unsatisfactory. We have also suggested that one source of the problem is the way in which the quest for demarcation criteria is mixed up with the problems of trying to understand science in retrospect. Another, related dimension to the problem is that the organization and conception of science has, at a general level, itself varied over time. In other words, the way in which science might be defined has itself varied in response to the organizational and social factors which bear upon its boundedness. In this section, therefore, we present an outline account of general organizational changes in what has counted as science.

The 1960s saw a rash of studies of the statistical growth of science. In particular, it was noted that the rate of growth of science was exponential. As de Solla Price was among the first to point out, the rapid rate of exponential scientific growth far outstripped the exponential growth of population and the growth of Gross National Product [9]. This period of study of science also saw the frequent popularity of claims that soon every man, woman and child in the country would be a scientist. (This remark has been attributed to a man called Boring.) The general mood is captured in the apparently startling claim that: 'eighty per cent of all scientists who ever lived are alive today'. (Although it should be noted that a characteristic of exponential growth is that some approximation to this latter statement has always been true.) Price correctly pointed out the implication that science could no longer continue to grow at the same rate, that we had reached saturation.

The interest for our purposes is in the way in which science was conceived and operationalized for the purposes of statistical measurement. In every case, the statistician retrospectively construed cultural activities as comparable with what was presently known as science. Growth patterns were taken to suggest a boom in scientific knowledge, but it is of course highly problematic to equate an increase in scientific activity with an increase in scientific knowledge. This in turn raised questions about the efficiency and wisdom of further investing in scientific activity. But what in any case counted as scientific activity. If we review changes in the social organization of science since the seventeenth century, it is evident that 'science' has undergone significant and substantial changes.

The social organization of science can be conceived in three broad phases: amateur, academic and professional [10]. In the *amateur* period (roughly between 1600 and 1800), science took place outside universities, the government and industry as we now know them. The participants were professional and financially independent men, whose major social role lay outside their interest in science, but who met informally. These amateurs gradually evolved methods of communication between them, the exchange of letters gradually gave way to, or was supplemented by, the development of scientific journals. Participants in these social networks of amateurs conceived of themselves as interested in 'natural philosophy' and specialization was rare. The *academic* phase (1800-1940) was characterized by the need for longer and more technical training of new members of the scientific community to cope with the growing body of scientific knowledge, by the need for adequate resources or positions to support the full-time monitoring of the burgeoning scientific literature and by the increasing specialization of practitioners. As a result, scientific work tended to become focused on basic research within universities. Increasingly, scientific careers were organized along specialist disciplinary lines, and part of a scientist's obligations was the training of other new members of the scientific community. Although science was increasingly supported by public funds, neither universities nor governments were allowed to interfere directly in the academic freedom of scientists. Consequently, the direction of scientific development was shaped almost entirely by the internal momentum of the scientific community. Since 1940, science has been rather more *professional* than academic. Although unplanned research has taken place within universities, scientific research has now become so expensive, especially in terms of capital investment, that only centrally located government funds can support it. Hence the increased interest and influence of non-scientific sponsors in the progress of science. Increasingly, the work of science is adjudged in terms of its value for economic prosperity and for security. Consistent with this growing emphasis upon the applicability and utility of science, is the gradual increase of scientific effort directly related to industrial interests: most major industrial firms have Research and Development laboratories on site. And the post-war regeneration of awareness of the relationship between science and the wider society has also restimulated concerns about the impact of science 'on society'.

ESSENTIALISM AND NOMINALISM

Our efforts to answer 'What is Science?' have yielded two senses in which science is highly variable. Not only have philosophers disagreed about the characteristics which distinguish science from other activities, but the character of science has been shown to be historically variable. There are two significantly different reactions to this variation. On the one hand, we could take this variability to result from the complexity of science itself. In other words, we might say, it is difficult to pin down the actual nature of science precisely because it is such a complex, shifting animal. Let us call this the *essentialist* position. In this view, science continues to be conceived as an object, a coherent entity or method, albeit one whose definition and description are difficult. *Importantly, this reaction neither abandons nor modifies in any significant way the view that there actually is something 'out there' called science.* This reaction merely postpones, as it were, the business of finding a definitive answer. By contrast, a *nominalist* reaction to variation in definitions of science takes the position that the quest for definition is ultimately futile. Attempts to specify one or other criteria of demarcation ignore what turns out to be an intriguing and fundamental characteristic of science, that it is constantly open to renegotiation and reclassification. From this point of view there is no such thing as 'science' or 'the scientific method' except that which is variously and multiply attributed to various practices and behaviours. 'What counts as science' varies according to the particular textual purposes for which this is an issue. Whereas essentialism tends to the view that definitions of science are, at least in part, a reflection of the characteristics of an actual (transcendental) object called 'science', nominalism suggests that features proposed as characteristic of science stem from the definitional practices of the participants (philosophers, historians and sociologists) themselves.

As we shall see in later chapters, this distinction is far more than a mere methodological observation about different approaches to the study of science. It underpins a basic dilemma for all social science which adopts a relativistic approach to the phenomenon it studies: to what extent do the features, characteristics and definitions of phenomena reflect the defining practices (constructional work) of the people involved rather than the 'actual character' of the phenomena? We shall also see that, although recent work in the social study of science demonstrates sympathy with the nominalist position, it remains ambivalent about the implications for its own research. This literature often concurs that it is not helpful to try and settle the question of what science comprises; instead, the importance of the idea of 'science' is in its use as a resource for characterizing the work and attitudes of others; and this opens the way to studying how the term 'science' is attributed to (or withheld from) various practices and claims. Unfortunately, however, this line of argument is compromised by the fact that the social study of science itself construes 'science' as an object for its own 'textual purposes'. The social study of science embraces the nominalist position with respect to the efforts of others to specify what counts as science, but tends to follow an essentialist line in its own practice. This is a significant problem, not just for the social study of science, but for all efforts to develop a critique of science.

It is important to elaborate the dominance of the essentialist reaction in some detail. It has been, and continues to be, an important influence on all attempts to come to terms with the phenonomenon of science. In the next sections, I outline the influence of the essentialist position in two areas of scholarship — the classical sociology of knowledge and the sociology of science — which provide the backcloth to the modern social study of science.

THE CLASSICAL SOCIOLOGY OF KNOWLEDGE

Sociological interest in a particular phenomenon is frequently justified in terms of its social relativity. This is the 'it could be otherwise' maxim which is crucial to all forms of relativistic social science. Sociological interest in knowledge is often justified in the same way. Thus, Pascal is said to have remarked that what is truth on one side of the Pyrenees is error on the other. The presumed variation in what counts as knowledge permits us to pose sociological questions about the source, extent and manner of these variations. As authors like Berger and Luckman point out, these questions are traditionally distinct from those of philosophers [11]. Whereas the latter are concerned to determine criteria of knowledge — they are in effect attempting to specify what should legitimately count as knowledge — sociological questions tend to be less interested in the status of knowledge. Sociologists seek instead merely to document different claims to legitimacy as a prelude to explaining the differences.

Social Context ——————— Human Thought/Knowledge

Specifically sociological (as opposed to psychological or economic, etc.) interest in knowledge arises in virtue of conceptualizing sources of variation in terms of social context or attributes. Variations in knowledge are thus associated with differences in class background, religious affiliation, 'social being', social context, social groups, society, culture, race and so on. The sociology of knowledge has encompassed kinds of human thought and knowledge as varied as legal, political, religious literary and artistic ideas. The curious, but much noted, exclusion from this latter list is science. Especially in the hands of the classical writers, the sociology of knowledge fought shy of attempting to explain scientific knowledge.

An obvious reason for the neglect of science stems directly from the way the sociology of knowledge is conceived. Science is exempt from sociological analysis because it is thought not to admit to the variation characteristic of other forms of knowledge. Science is assumed to be the form of knowledge which, *par excellence*, remains unaffected by changes in social context, culture and so on. The recent social study of science challenges just this assumption. The argument now is that the universalism of scientific truths is a myth, that the appearance of universalism is the *upshot of* (that is, the consensual response to) a complex social process whereby variations in the form and legitimacy of scientific knowledge claims are gradually eliminated. The apparent lack of social variation in scientific knowledge is the accomplishment of, not the condition for, science.

In broad terms, Marx's formula for a sociology of knowledge was that man's social being determines his thought and consciousness [12]. By social being, Marx interpreted the 'social context' side of the equation in terms of class position, most notably, man's relation to the means of production.

The existence of revolutionary ideas presupposes the existence of a revolutionary class. False consciousness is the product of circumstances in which one social class adopts the thought appropriate to (and fostered by) another; the working class thus displays false consciousness when it adopts the ideology of the owners of the means of production. Although Marx subsequently modified his initial insistence on the social determination of ideas (an early reaction to the idealism of previous writers), his sociology of knowledge never became a fully developed part of his writing. His contribution in the area was subsumed under the dominant interest of analysing the conditions for social change. In particular, he was concerned with discerning the origins of *false* knowledge, whereby the revolutionary potential of the working class was concealed from themselves. His relative disregard of science echoes this concentration on sources of distortion.

Mannheim attempted to transform the Marxian approach into a more general tool for the sociological analysis of knowledge [13]. In particular, he was concerned to extend both rubrics ('social context' and 'human thought') to include a greater diversity of variables. Marx's predominant interest in the association between material (class) interests and intellectual attitudes thus gives way to, say, the connection between the intellectual motivation of a social group and that group's style of thought. Rather like Weber, Mannheim insisted on extending the range of categories subsumed under 'social context' by Marx. In Mannheim's work, status, group membership and social role are all potential determinants of knowledge. More importantly, Mannheim differs from Marx in his wish to extend the analysis to all ideas including those held to be true. Marx's concentration on ideology as a source of distortion (based on the assumption that only proletarian classes can achieve true knowledge) thus gave way to the view that all ideas are ideology: truth can only be said to exist within the specific world-view of its adherents.

Mannheim's sociology of knowledge, albeit largely programmatic rather than empirical, was thus more radical (epistemologically, if not politically) than that of Marx. In particular, it opened the way for the sociological analysis of knowledge systems regarded as reliable producers of truth. But despite these arguments against the partial scope of the sociology of knowledge, Mannheim failed to press his scheme to the understanding of science and mathematics. Mannheim's Mistake was to stop short of their analysis despite his principled argument for the general applicability of the sociology of knowledge.

Durkheim, the last of the sociology of knowledge triumvirate considered here, further broadened the rubrics in the sociology of knowledge equation [14]. He brought to bear an anthropologically inclined perspective on such aspects of human thought as morals, values, religious ideas, basic forms of classification, fundamental categories of human thought, space and time. Like the very idea of society, these kinds of knowledge, belief and ideas are part of the collective consciousness. They cannot exist independently of man's social existence. So the kinds of knowledge and belief held by man enjoy a kind of isomorphism with the form of society he produces and sustains. For example, religion is a belief system in terms of which men organize their lives and categorize their world (distinguishing, for example, between the sacred and the profane). At the same time, religion is a 'social fact' which provides a source of constraint upon social activity and behaviour. Objects are classified in societies in a way which reflects and extends existing social classifications. Thus, ideas about space reflect the basic material and social organization of society; time divisions mirror the ways in which rituals and feasts are organized.

Durkheim thus establishes an intriguing anthropological framework for the study of science. The isomorphism (or parallelism) between social and physical worlds suggests we can understand the structure of knowledge about the physical world as a reflection of structure in the social world; our apprehension of nature will display the organization and arrangement of our social institutions. Unfortunately, it seems that Durkheim argued himself out of this fascinating possibility. His overriding concern with the evolution of societies — the move from forms of mechanical to organic solidarity and beyond — led him to regard science as knowledge of a different status from all other kinds. Science replaced religion, not just as a result of basic changes in forms of social organization, but as an evolutionary advance which severed the ties between social organization and intellectual activity. For Durkheim, science was important as a form of knowledge which, unlike any other, escaped its social context. Durkheim thus reaches the position that science is a special case, exempt from treatment in the general anthropological perspective with which he began.

We see that Durkheim, like Marx and Mannheim, excluded science from the sociology of knowledge on the grounds that science is a special case [15]. In effect, the assumption of these authors is that there is indeed something specific to science which sets it apart from other kinds of knowledge. In line with the essentialist reaction to the demarcation question, these authors assume the special character of science without specifying what that might be. To the extent that these authors and their contributions to social science continue to be influential, it is this traditional position with which the new social study of science has had to contend.

THE SOCIOLOGY OF SCIENCE

At the same time, the new social study of science has had to contend with a largely separate tradition within sociology — the sociology of science —

which has also adopted an essentialist position on the character of science.

Our earlier historical overview of the emergence of science as social institution established a source of variation in conceptions of science quite distinct from problems of demarcation. 'Science' was said to be done in the seventeenth century as gentleman amateur generalists exchanged letters and met informally; but 'science' was also said to be done by post-war industrialist specialists with access to highly sophisticated electronic communication facilities. It is tempting to conclude that science has become 'more social': greater specialization and differentiation has demanded increased social organization and control (both internal and external); much science is capital intensive — huge investment in equipment and specialist techniques has encouraged teamwork. The days of the (relatively) isolated individual scientist have been replaced by her situation within a complex social network and her becoming subject to a series of social forces and pressures. The scientist now belongs to a definite and often close-knit social group. Scientist's relationships with one another are defined by what counts as being scientific. Thus, when science requires training we find a series of social roles and status relationships associated with teaching and being taught. Social isolation is further decreased as the scientist emerges from the confines of academia and is made responsive to the demands of industry and government. Although it is possible to make a case that the wider values and beliefs of society impinged upon the gentleman amateur, we now seem to have a more immediate social influence: the individual scientist is part of an institutionalized social system [16].

It is important to recognise, however, that this interpretation of changes in science deploys a limited and specific sense of social. In particular, this usage of 'social' tends to emphasize those circumstances and effects which lie *external to* the intellectual activity of the scientist. This is consistent with the essentialist position: the actual character of science (in particular, the esoteric details of the content of scientific knowledge) is treated as independent of (prior to) and separate from its practitioners. Against this, it can be argued that activities like interpreting, proving, marshalling evidence and making observations have always been 'social' in the more phenomenological sense of the term. Thus, the isolated individual scientist is inextricably immersed in a 'language game' whether she is part of the seventeenth or twentieth centuries. She engages with the meaning of her actions (for example, the words she is writing) and she apprehends the possible reaction to and treatment of them, their persuasiveness and so on, on the basis of her membership of a language community.

Unfortunately, this phenomenological perspective was for a long time neglected in favour of attention to the institutional/structural sense in which scientific action is social. Thus, at the hands of the sociology of science, especially as practised by followers of Robert Merton, the central concern was the manner in which science, as a fast-growing social institution, organized and regulated itself [17]. Much emphasis was given, to the nature of the relationship between knowledge producers: their social roles, the nature of the reward system, competition and, especially, the system of norms which guided scientists' actions. As is now well recognized, this concentration on relationships between *scientists* proceeded at the expense of attention to the ways in which different kinds of *scientific knowledge* came to be produced and accredited [18]. The sociology of science thus adopted an essentialist line by presuming the actual character of science to be beyond its province of inquiry.

CONCLUSION

We have seen that current conceptions of science and attempts at its social analysis take place against a rich mosaic of traditions in the history, philosophy and sociology of science. The core assumptions of these traditions provide both a springboard for, and a series of constraints upon, our understanding of science. In sum, the main constraints are:

(1) The persistent idea that science is something special and distinct from other forms of cultural and social activity, despite the major disagreements and changes in the opinions of philosophers attempting to elucidate criteria of distinctiveness. Instead of treating them as rhetorical accomplishments, many analysts continue to respect the boundaries which delineate science from non-science. Many other analysts explicitly eschew the possibility of demarcation but nonetheless continue to cast their discussion in terms of boundaries. The continued use of a format which construes science as an object tends to reinforce, rather than challenge, the distinctiveness of science.

(2) The persistence of what has been called the 'received' (or standard) view of science. This view includes the assumption that the objects of the natural world are real, objective and enjoy an independent pre-existence. A corollary is that the social origins of scientific knowledge are almost completely irrelevant to its content. Scientific knowledge, in this view, is not amenable to sociological analysis, simply because it is its own explanation: scientific knowledge is determined by the actual nature of the physical world.

(3) The persistent notion of knowledge as an individualistic and mentalistic activity; the enduring respect for the work and achievements of 'great

men'. This notion stems from and, in turn, reinforces the idea that agency is basically incidental to the actual objective character of the natural world 'out there'. Prevalent public images of science underscore this view. It is notable, for example, that both science journalists and the media more generally have almost wholly failed to take up the relativist themes which inform the recent sociology of scientific knowledge. Instead, news about the production of science continues to emphasize the heroic feats of individuals.

(4) The unwillingness to face the radical consequences for one's own efforts of a critical attack on science. This is the 'problem' of reflexivity, about which we shall have much to say.

Common to each of these barriers is the commitment to essentialism. Over and above the question of whether or not science differs from non-science, each of these prejudices shares the view of science as a concrete, identifiable activity. The constraints upon our understanding of science thus arise because various traditions of scholarship have construed science as an *object*, upon which they have brought to bear their own toolkit (or set of concepts).

However, the nominalist reaction to the problem of demarcation asks that we dig a little deeper. In particular, it suggests that we look critically at the very idea of 'investigating an object' which permeates traditional scholarship. As we shall see, some approaches to the study of science, notably the 'new' sociology of scientific knowledge (SSK), have enjoyed some success in escaping their origins. But they badly need to shake off a further enduring preconception before being able to free themselves in any significant way from the confines of the past. This preconception — which we treat in terms of 'representation' — is the topic of the next chapter.

NOTES

[1] In 1984 the Social Science Research Council became the Economic and Social Research Council.

[2] See, for example, R. Wallis, 'Science & Pseudo-Science', Social Science Information, Vol. 24 (1985), pp. 585–601.

[3] K. R. Popper, *The Logic of Scientific Discovery*, Hutchinson, London (1980) and *Conjectures and Refutations*, Routledge & Kegan Paul, London (1963).

[4] I. Lakatos, 'Falsification and the Methodology of Scientific Research Programmes' in I. Lakatos and A. Musgrave (eds.) *Criticism and the Growth of Knowledge*, Cambridge University Press, Cambridge (1974) pp. 91–196.

[5] T. S. Kuhn, *The Structure of Scientific Revolutions*, 2nd edition, University of Chicago Press, Chicago (1970).

[6] Wallis, op. cit. (note 2).

[7] S. Shapin, 'The Politics of Observation: cerebral anatomy and social interests in the Edinburgh phrenology disputes' in R. Wallis (ed.) *On the Margins of Science: the Social Construction of Rejected Knowledge*, Sociological Review Monograph 27, Keele University (1979), pp. 139–178.

[8] H. Collins and T. Pinch, 'The Construction of the Paranormal: nothing unscientific is happening' in Wallis, op. cit (note 7), pp. 237–270; *Frames of Meaning: the social construction of extraordinary science*, Routledge & Kegan Paul, London (1982).

[9] D. J. de Solla Price, *Little Science, Big Science*, Columbia University Press, New York (1963).

[10] See, for example, H. Butterfield, *The Orgins of Modern Science*, G. Bell, London (1968); H. Rose and S. Rose, *Science and Society*, Penguin, Harmondsworth (1969).

[11] P. L. Berger and T. Luckmann, *The Social Construction of Reality*, Penguin, Harmondsworth (1967), pp. 13–15.

[12] K. Marx and F. Engels, *The German Ideology*, edited by R. Pascal, International Publishers Inc., New York (1963).

[13] K. Mannheim, *Ideology and Utopia*, translated by L. Wirth and E. Shils, Harvest Books, New York (1936).

[14] E. Durkheim, *Elementary Forms of Religious Life*, George Allen & Unwin, London (1915).

[15] Cf. M. J. Mulkay, *Science and the Sociology of Knowledge*, Allen & Unwin, London (1979).

[16] Rose and Rose, op. cit. (note 10).

[17] R. K. Merton, *The Sociology of Science: Theoretical and Empirical Investigations*, University of Chicago Press, Chicago (1973). This volume collects Merton's major writings on sociology of science. Examples of main publications which by and large fall in with the Mertonian programme include: J. Ben-David, *The Scientist's Role in Society: a comparative study*, Englewood Cliffs, New Jersey, Prentice-Hall (1971); J. Cole and S. Cole, *The Social Stratification System in Science*, Chicago University Press, Chicago (1973); J. Gaston, *Originality and Competition in Science: a study of the British High Energy Physics Community*, University of Chicago Press, Chicago (1973); W. O. Hagstrom, *The Scientific Community*, Basic Books, New York (1965); N. Storer, *The Social System of Science*, Holt, Rinehart & Winston, New York (1966); H. A. Zuckerman, *Scientific Elite: studies*

of Nobel laureates in the United States, University of Chicago Press, Chicago (1974).

[18] Some early critiques of the Mertonian school of the sociology of science are to be found in B. Barnes and R. G. A. Dolby, 'The Scientific Ethos: a deviant viewpoint', *European Journal of Sociology,* **11** (1970) pp. 3–25; M. J. Mulkay, 'Some Aspects of Cultural Growth in the Natural Sciences', *Social Research,* **36** (1969). See also B. Latour and S. Woolgar, *Laboratory Life: the construction of scientific facts*, 2nd edition, Princeton University Press, Princeton (1986), Chapter 5.

2

Representation and the Methodological Horrors

Science would be superfluous if there were no difference between the appearance of things and their essence. [1]

It can not be that the axioms of argumentation suffice for the discovery of great works, since nature is more complicated many times over than argumentation. [2]

The excitement of current work in the social study of science is that it is struggling with a hundred years of tradition. Its significance lies not just in providing more or different news 'about science', but in its potential for reevaluating fundamental assumptions of modern thought. In the last chapter we identified essentialism — the idea that discrete objects exist independent of our perception of them — as one of the more significant constraints which this tradition imposes upon our efforts to understand science. In this chapter, our objective is to explore the implications of essentialism and to formulate some policies for resisting or, at least, making us aware of, these constraints.

The standpoint of essentialism finds support in the idea of representation. Representation is the means by which we generate images (reflections, representations, reports) of the object 'out there'. Representation is axiomatic not just to science but to all practices which trade upon an objectivist epistemology, in short, to all activities which claim to capture some feature beyond the activity itself. Of particular importance, this means that representation sustains not only science but also the attempts of social scientists and others to analyse science.

Discussions in and about science are characterized by their reliance on a fundamental dualism — the supposed distinction between 'representation' and 'object'. In fact this is just one of a number of ways of expressing this dualism. For example:

representation	object
image	reality
document	underlying pattern
signifier	signified
action or behaviour	intention
action or behaviour	cause
language	meaning
explanandum	explanans
knowledge	facts

More specific examples, taken from a range of sciences and social sciences, might include:

voltmeter reading	voltage
pen chart recorder	changes in resistance
questionnaire response	respondent's attitude
what was said	what was meant
documentary evidence	the historical situation
gesture	meaning or intention
photograph	photographed scene

This scheme is perfectly general and the range of examples vastly extendable, demonstrating the pervasiveness of the notion of a dualism between a representation and its object.

THE PROBLEM OF REPRESENTATION

This scheme of dualities is the basis for a Problem which will not go away: how can we be sure that the left-hand side (representation) is indeed a proper, true reflection of the right-hand side (object) [3]? It is a Problem of methodological adequacy: what grounds provide the warrant for the relationship between the objects of study and statements made about those statements? The Problem is as widespread as the dualism itself. Thus, in principle, this problem is equally applicable to the work of both social and natural sciences. For example, in the work of a solid state physicist, it might appear as a concern for the correspondence between the inscription of a pen chart recorder and the current state of atomic alignment in a sample of metal alloy. For the sociologist, the problem is classically associated with the connection between a particular social indicator and the corresponding social reality. Among historians, we find a concern for what about a particular historical situation is revealed by particular items of documentary evidence. A vast number of similar examples could be given, suggesting that, in principle, the Problem pervades every aspect of research practice or inquiry more generally.

THE METHODOLOGICAL HORRORS

For any particular representation–object couple, the Problem of the adequacy of connection between representation and represented object can rear its head in any (or all) of three main ways. These constitute the 'methodological horrors', an inventory of the ways in which attempts to effect connections between representation and object can go wrong. Here we describe these 'horrors' using the terminology of early ethnomethodology [4].

(1) Indexicality
The link between representation and object is indexical. In some formulations, the 'meaning' of the document is said to be indexical. In other words, the underlying reality of a representation is never fixed and always able to change with occasion of use. This means that it is not in principle possible to establish an invariant meaning for any given representation; any given sign (document) can in principle be taken as indicative of at least two alternative underlying realities (objects, meanings). It is thus always possible to nominate an alternative to any specific proposed meaning. The constant availability of alternative versions of the same event has the fairly obvious consequence that *all* attempts to do representation (specify the meaning, describe the object, nominate the cause, and so on) are defeasible (that is, are capable of being defeated).

(2) Inconcludability
The task of exhaustively and precisely defining the underlying pattern (meaning) of any one representation is in principle endless. In other words, it is always possible to ask for further clarification, elaboration, elucidation and the like. Attempts to meet this request are ultimately doomed to failure in the sense that they inevitably involve the use of other representations (in the form of words, signs, gestures, graphs, and so on) as part of the clarification process and these documents can themselves be subject to the same kinds of request for yet further elucidation. It is impossible to provide an exhaustive account of (say) the meaning of a representation since all accounts comprise meanings which remain unexplicated; the very nature of the task of exhaustive explication thus multiplies the features of the task.

(3) Reflexivity

The relationship between representation and represented object is reflexive in the particular sense due to Garfinkel [5]. That is, the intimate interdependence between representation and represented object is such that the sense of the former is elaborated by drawing on 'knowledge of' the latter, and knowledge of the latter is elaborated by what is known about the former. The character of the representation, as perceived by the actor, changes to accommodate the perceived nature of the underlying reality and the latter simultaneously changes to accommodate the former. The establishment of a connection between representation and object is thus a back-and-forth process. In Garfinkel's terminology, 'members' accounts are constituent features of the settings they make observable' [6]. This means, in particular, that it is not possible to conceive of component parts of any representation–object couple as straightforwardly independent. The consequences for certain forms of interpretive practice are profound. For example, in models of causal explanation, the horror of reflexivity suggests we recognize explanans and explanandum as intimately and inextricably intertwined.

MANAGING THE METHODOLOGICAL HORRORS

The Problem of representation refuses to go away in the sense that although it can be 'dealt with' or 'managed' on any specific occasion, the general form of the difficulty remains and threatens to re-emerge at each next instance of interpretation (representation). The problem is a general and irresolvable problem of epistemology, which requires management whenever it makes its appearance, lest it entirely disrupt research practice. The magnitude of the Problem is horrendous in that it applies in principle not just to certain practices on certain occasions of academic research, but to any act of interpretation whatsoever. It is possible to identify four main kinds of strategy which are used to manage the methodological horrors. In various ways, each strategy seeks to deny, evade or minimize the Problem. Common to each is the effort to negate the generality (and stress the particularity) of the Problem.

Strategy 1: Appeal to an hierarchy of knowledge

A first management strategy denies the generality of the Problem by appealing to a presumed hierarchy of situations and occasions in which the Problem has differential applicability. For example, it is said that as a practical matter we observe that certain kinds of representation are less dependable than others: some connections between representation and object are deemed more problematic than others. For example, the psychoanalyst has much greater difficulty in establishing the meaning of dreams (that is, in demonstrating the correspondence between dreams and reality) than does the astrophysicist in demonstrating the pattern underlying the clustering of galaxies. The last set of examples (above, page 31) of the representation–object couple are arranged, in a rough and ready fashion, in descending order of perceived reliability. This reflects the view that say, the correspondence between voltmeter readings and voltage (in physics) is more reliable than between an interview response and the respondent's meaning or attitude (in social science).

This strategy thus seeks to deny that the Problem applies equally to all kinds of interpretations. It is important to be clear, however, that this perception of relative reliability is the *consequence* of natural science's claimed superiority in establishing correspondence, not its *cause*. It is in virtue of accepting the claim of science to be able to effect superior correspondence that we take voltmeter readings to be more reliable. The appeal to a hierarchy of knowledge of different degrees of reliability effectively begs our main question. The central objective of the social study of science is to determine what features of science bring about the perception of its superiority.

Strategy 2:   Construing the Problem as a merely 'technical' difficulty

A second management strategy is to transform the Problem into a merely technical difficulty. For example, the problematic relationship between the shape of the output trace and changes in atomic alignment are attended to in terms of factors (the speed of the trace, the setting of the gain amplifier, the polarity of the electrical connections and so on) which might have biased the pen recording. The basic assumption is that the removal or rectification of such factors will eventually (perhaps only after much effort) restore the rightful connection between trace and atomic alignment. Fundamental, omnipresent (methodological) problems are thus treated as mere 'technical' (methods) difficulties which occasionally come to the fore by virtue (say) of members' use of faulty interpretive procedures. But the general form of the Problem still lurks: on what grounds can we establish a connection between any of these 'biasing factors' and the behaviour of the trace? In practice, such considerations are usually relegated to the background by attaching primary significance to the practical resolution of the initial difficulty. Thus the Problem does make its presence known, but it is managed by transforming it into a resolvable and limited matter of technical adequacy. Sometimes this way of managing the problem is accompanied by statements to the effect that preoccupation with technical matters is an overindulgence which detracts from the research goals at hand.

Strategy 3: Denying the significance of the Problem

A third management strategy comprises the suggestion that the whole basis for the Problem is artificial, that it is a problem of no consequence. According to this line of argument, the articulation of the Problem is an artificial exercise in philosophy; the fact that all connections between representation and object can be shown *in principle* to be defeasible is of no consequence since people do not *in practice* concern themselves with such difficulties; since people are not routinely conscious of the methodological horrors, it makes no sense to interpret their actions as if they were deliberately seeking to avoid them.

But the suggestion that people need to be conscious of the methodological horrors is misleading. The horrors exist as a form of argument which has the potential to be invoked on any occasion. They clearly could be part of 'merely' philosophical conjecture. But the fact that they *are* invoked in practice is evident from instances of controversy between scientists (for example, about the inadequacy of a competitor's results) and, more significantly, from the invocation of relativism in many forms of social science argument. This use of relativism posits that representations (definitions, labels, interpretations) could be other than they are. For example, the 'deviant' may not in fact be deviant; in other times (places, societies) she could (would) be labelled differently. This kind of relativism amounts to the sort of invocation of indexicality associated with our common-sense notion that 'things may be other than they seem'.

Strategy 4: Construing the Problem as a problem for others

Finally, an important and widespread management strategy entails the assumption, either explicit or *de facto,* that although the horrors might arise in the work of others, they do not intrude upon one's own work. This involves a subtle feature of argumentative discourse such that the fallibility of one's own argument is de-emphasized, while the fallibility of the argument of one's subjects is highlighted. Typically the author (researcher) proceeds as if she is acting at a more robust level of representational practice than the subjects (objects) being studied. The Problem is portrayed as essentially only a Problem for others.

This is of particular significance when we recognize that the business of constructing a text (that is, writing) is no less immune to the methodological horrors than, say, the work of interpreting (constructing) rock formations in Devonian geology [7]. The management strategy is to proceed as if the representations which we, as social analysts of science, produce are somehow less open to Problems than the representations produced by the scientists we study. We shall return to a longer discussion of this particular management strategy in Chapter 5.

A further notable feature of this particular strategy is its use in generating formulae for social science analyses which espouse a form of relativism. In a manner similar to that just noted (under Strategy 3), relativist analyses of social phenomena invoke the methodological horrors with respect to certain social phenomena in order to open them up to investigation. For example, the notion (indexicality) that the meaning of a document changes with the occasion of its use, is taken as the basis for research programmes which investigate patterns of correspondence between 'context' and 'meaning'. Crucially, however, the same notion is *not* applied to the documents produced by (the results of) the research programme; the Problem is thus construed as a problem for the subjects but not for the researchers [8].

CONCLUSION: TWO POLICIES FOR THE STUDY OF SCIENCE

We shall see, especially in Chapter 4, how the discourse of science — and by implication the discourses of all forms of inquiry with scientific pretensions — is organized to reinforce the ideology of representation. We have already suggested that the effect of upholding the idea of representation is to deny the generality of the methodological horrors. It is clear, however, that a critical appraisal of the idea of science must challenge the very idea of representation. In particular, we need to realize the extent to which our own efforts (as social scientists) are themselves beholden to the ideology of representation.

In order to begin to meet this objective and to tackle the critical problems emerging from the heart of our representation couple, we formulate two main policies: *inversion* and *feedbacking*.

(1)   Inversion

Our first policy is to be critical of relationships which construe a unidirectional connection between two elements of the representation couple. We need to take issue both with the idea that elements of the couple are distinct and with the notion that the object is prior (or antecedent) to the representation. Inversion asks that we consider representation as *preceding* the represented object. In Chapter 3, for example, we consider the value of arguing that discovered objects are constituted through, rather than revealed by, their discovery. Inversion is thus a direct challenge to the position referred to in Chapter 1 as essentialism.

(2)   Feedbacking

Our second, related, policy is to emphasize rather than suppress the rebounding connections between 'science' as an object and our attempts to produce a study 'of' science. The aim of feedbacking is to resist the persistent construal of science as a distinct topic for study, an object 'out there', beyond us *qua* observers/inquirers, and essentially separate and distinct from our own writing practices. Since science is quintessentially engaged in the achievement of order, of accomplishing stable relationships between the world of objects and the world of representations, our policy is to ask what can be done to disrupt this. What kind of inquiry would generate instability in respect of the dualism of the representation couple? In addressing this question we aim to develop a better understanding of the core features which sustain the idea of science.

With these two policies at hand, we now begin our critical examination of the achievements of the recent social study of science.

NOTES

[1] A remark attributed to Karl Marx in E. Gellner, *Cause and Meaning in the Social Sciences,* Routledge & Kegan Paul, London (1973).
[2] A remark attributed to Francis Bacon in an textbook on quantitative methods in social science.
[3] S. Woolgar, 'Irony in the Social Study of Science' in K. D. Knorr-Cetina and M. Mulkay (eds.) *Science Observed: perspectives on the social study of science,* London, Sage (1983), pp. 239–266.
[4] H. Garfinkel, *Studies in Ethnomethodology,* Englewood Cliffs, New Jersey, Prentice-Hall (1967). It should be noted, however, that few of the later interpreters of Garfinkel would wish to pose the issue in terms of 'horrors'. With the possible exception of writers such as Blum and McHugh (A. Blum, *Theorising,* Heinemann, London (1974); P. McHugh, S. Raffel, D. Foss and A. Blum, *On The Beginning of Social Inquiry,* Routledge & Kegan Paul, London (1974)), most ethnomethodologists have tended to play down the scepticism of Garfinkel's writings, in favour of developing an empirical research programme. See, for example, W. Sharrock and B. Anderson, *The Ethnomethodologists,* Ellis Horwood, Chichester/Tavistock, London (1986) and J. Heritage, *Garfinkel and Ethnomethodology,* Polity Press, Cambridge (1984).
[5] Garfinkel (1967), op. cit. (note 3).
[6] Ibid., p. 8.
[7] M. J. S. Rudwick, *The Great Devonian Controversy: The Shaping of Scientific Knowledge Among Gentlemanly Specialists,* University of Chicago Press, Chicago (1985).
[8] S. Woolgar, 'Irony in the Social Study of Science', in K. D. Knorr-

Cetina and M. Mulkay (eds.), *Science Observed: perspectives on the social study of science,* Sage, London (1983), pp. 239–266; S. Woolgar and D. Pawluch, 'Ontological Gerrymandering: the anatomy of social problems explanations', *Social Problems,* **32** (1985), pp. 214–227.

3

Opening the Black Box:
Logic, Reason and Rules

Thus far we have discussed the wide variation in conceptions of science (Chapter 1) and noted the importance of the idea of representation (Chapter 2) which, we suggested, is deeply implicated in all efforts to be 'scientific'. In this chapter we start to look at how sociologists (in particular, but also some historians and philosophers) have attempted to escape the constraints upon our understanding of science exerted by the dominance of essentialism. As a specific example we examine attempts by sociologists to study logic or reasoning, especially that said to be involved in promoting scientific and mathematical laws and truths.

One of the major consequences of sociologies of science which adopted the received view was their neglect of the nature and structure of scientific knowledge. In general, these sociologies operated by treating scientific knowledge — the theories, formulae, physical laws, mathematical equations and proofs — as a black box. For sociological purposes it was assumed that nothing was to be gained by opening the box and scrutinizing its contents; the social origins of scientific knowledge were regarded as simply not relevant to its content. Some sociologists even went so far as to suggest that attention to content detracts from the proper task of sociological analysis. In terms of our representation couple, this attitude treated the relationship between 'scientific knowledge' and 'the objective, natural world' as a black box. The nature of the relationship was considered beyond the realm of sociological inquiry: it was neither necessary nor desirable to consider how they are conjoined.

A particular feature of this viewpoint was that the adequacy of the connection, whether or not scientific knowledge is an accurate representation of the world, was regarded as a question of methodology rather than sociology. Thus Merton, when introducing his famous discussion of the normative ethos in science: 'To be sure, methodological canons are often both technical expedients and moral complusives, but it is solely the latter which is our concern here. This is an essay in the sociology of science not an excursion into methodology.' [1]. In this usage, Merton portrays 'methodology' pejoratively, as mere methodology. Moral compulsives are said to be socially organized and structured but technical expedients (methodology) are assumed to be socially neutral. When incorrect knowledge of the world arises, the source of error is a misapplication of method, not a question for sociology. In this case, the 'distorting factors' in Fig. 3.1 (which deflect the

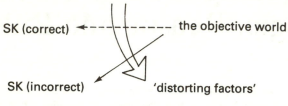

SK (correct) ◄─ ─ ─ ─┼─ ─ ─ ─ ─ the objective world

SK (incorrect)   'distorting factors'

Fig. 3.1.

rightful connection between world and scientific knowledge) correspond to the misapplication of scientific method.

Within the received view, some sociologists and historians have considered methodology as a social phenomenon, but have confined their attention to occasions resulting in incorrect scientific knowledge. This approach has been called the sociology of error. Sociologists are only consulted when things go wrong, in cases of deviation from the supposedly true path between the world and knowledge about it. Their brief is then to discern the source and nature of (external) social factors which lead to a distortion of knowledge about the world. Typically, in these sociological accounts, competition for rewards (and resources) appears as the cause of a deflected connection between the world and true knowledge. This position, it should be noted, is an asymmetric sociology of scientific knowledge: social factors are relevant in cases of false or incorrect knowledge; but the sociologist has no part to play when methodological connections are effected so as to produce correct knowledge. The sociology of error investigates the production of 'incorrect' knowledge but otherwise treats the generation of scientific knowledge as a black box.

The crucial flaw in sociologies of science which adopt the received view is their uncritical acceptance of what is said to count as 'true' and 'false' knowledge. Where knowledge was held to be correct, they saw no necessity for their involvement; where knowledge was deemed incorrect, they took this assessment as their starting point to ask what could have lead scientists to go wrong. They failed to consider that the determination (definition, assessment) of the truth status of knowledge is itself a social process. It was the major achievement of writers like Kuhn to establish the historically (and, by extension, the socially and culturally) relative character of scientific truths [2]. Consequently, the sociologist could no longer accept as given the distinction between true and false scientific beliefs. Instead, the sociological task was to discern what *counts* as true and false belief; in particular, what social processes are involved in the construction, assessment and evaluation of knowledge. It became clear that a sociological understanding of the construction of scientific knowledge requires a sophisticated appreciation of the technical content of the knowledge at issue and, preferably, a close contemporaneous investigation of scientists' technical endeavours (see Chapter 6).

The received view is consistent with sociologists' inattention to the content of scientific knowledge, an emphasis on science as a social institution and on the social relationships between knowledge producers. For this reason, the style of (traditional) sociology of science which adopted the received view was essentially a *sociology of scientists*. By contrast, more recent work emphasizes the relativity of scientific truth, calls for a sociological analysis of technical content and thereby attempts to open the black box of scientific knowledge construction. This latter style of sociology of science is more aptly dubbed a *sociology of scientific knowledge* (SSK).

In opening the black box, SSK significantly counters that aspect of the received view which sees no part for sociological investigation of the generation of scientific knowledge. But we recall (from Chapter 1) that the received view also implies a commitment to essentialism. To what extent does SSK rid us of this particular, important vestige of the traditional view of science? In order to tackle this question, we look in detail at one particular style of SSK.

THE CALL FOR A STRONG PROGRAMME IN THE SOCIOLOGY OF SCIENTIFIC KNOWLEDGE

The received view is consistent with some philosophical positions (which we shall refer to as 'rationalism') that the generation of true, correct knowledge simply does not require sociological explanation. From the perspective of rationalism, true, correct knowledge is explicable in terms of its rational merits: knowledge is truth which is believed for the right reasons. The general acceptance of the received view was for a long time reflected in the division of labour in the study of science — between sociologists investigating social factors affecting the production of erroneous knowledge and (rationalist) philosophers looking at the rational basis for truth. Philosophers and sociologists coexisted peacefully because each addressed the question — what is science — to different phenomena.

The situation changed dramatically with the insistence of sociologists that both truth and error are equally amenable to sociological analysis. David Bloor, in particular, challenged the exclusion of sociologists from studying how 'true' scientific knowledge is produced [3]. Whether labelled as 'true' or 'false', scientific knowledge could and should be the target of sociological analysis. Bloor complained that the insistence of rationalist philosophy on the inherently (given) true or false character of knowledge was directly opposed to attempts to study the social determination of 'truth' and 'falsehood'. Rationalist philosophy assumed that genuine (proven) knowledge was not *caused* (except, perhaps, by 'rational reasons'); it was the upshot of rational method and/or logical extrapolation from existing knowledge. In this view, logic, rationality and truth are their own explanation, and 'causes' are those (external) sociological, psychological and other factors which come into play in the genesis of false or erroneous knowledge: false knowledge is caused but true knowledge is simply the upshot of rational process.

Of course, it is just this view, in its popularized form which makes the idea of a sociology of scientific knowledge seem counter-intuitive: the sociology of the family, deviance, education and so on seem eminently possible, but what 'social' factors can possibly be involved in scientific knowledge? Are sociologists going to tell us that '2 + 2 = 4' is a social construct? The widespread feeling of counter-intuition associated with SSK is itself testimony to the influence of rationalist ideas. Scientific knowledge is assumed, by definition, to be precisely that which is *not* social; knowledge is thought only to become scientific in virtue of the exclusion of social factors.

Bloor [4] formulated four key requirements for a 'strong programme' which would take sociological study beyond the rationalist view of how scientific knowledge is generated: (1) Causality. The aim of the sociology of scientific knowledge is to discern which conditions bring about beliefs or states of knowledge. Bloor noted that these conditions could be psychological, economic, political or historical as well as social. (2) Impartiality. The sociology of scientific knowledge should not select instances for study with respect to their perceived truth or falsity, rationality or irrationality, success or failure. The emphasis is on the fact that truth, falsity and so on are perceived as such. These determinations are the upshot of social process and therefore part of the phenomenon to be studied. (3) Symmetry. Similarily, once instances of scientific knowledge have been chosen for study, the sociologist should use the same types of cause in explaining instances of scientific knowledge, whether they are classified as false or true, etc. In particular, argues Bloor, the sociologist should not invoke, say, sociological causes for 'false' beliefs but resort to, say, psychological (or worse, rationalist) causes for 'true' beliefs. (4) Reflexivity. In principle, the patterns of explanation of the sociology of scientific knowledge have to be applicable to sociology itself.

It should be clear from these tenets that mathematical statements such as '2 + 2 = 4' are as much a legitimate target of sociological questioning as any other item of knowledge (some sociologists use the term 'knowledge claim' rather than 'knowledge' to emphasize their impartiality). What kinds of historical conditions gave this expression currency and, in particular, what established (and now sustains) it as a belief? This kind of question is posed without regard for the (actual) truth status of the statement, but asks instead under what conditions it is regarded as 'true'. We are reminded, for example, that prior to the invention of mathematics, or for young children, such statements are meaningless; or again, that in the context of vector addition '2 + 2 = $\sqrt{2}$' is true. The strong programme claims not to be in any way assessing or evaluating the claimed truth status of the statement, but it is easy to see how analytic attention (of any kind, sociological or otherwise) to widely believed statements can be understood as casting aspersions on the veracity of the statement. Despite programmatic declarations of impartiality, the advocates of the strong programme are unavoidably embroiled in an agonistic discourse.

DEBATE OVER THE STRONG PROGRAMME

Controversy over the strong programme in the sociology of scientific knowledge has been heated: much vitriol and flying fur has accompanied exchanges between supporters of the strong programme and rationalist philosophers. The mood is captured by Bloor's infamous remark that 'to ask questions of the sort which philosophers address to themselves is to paralyse the mind' [5]. One possible (albeit *weak*) explanation for the intensity of the controversy emphasizes the importance of the entrenched division of labour between sociologists and philosophers of science. The call for a strong programme transgressed this division of labour by suggesting that the very content of scientific knowledge is amenable to sociological analysis; rationalist philosophy was outraged at the invasion of territory previously their exclusive domain.

Although the debate is about the best ways of finding out about the nature of science, its most curious feature is that participants themselves draw upon preconceived ideas about science in order to criticize or defend aspects of the proposed (strong) programme.

Bloor says that the principles of the strong programme 'embody the values which are taken for granted *in other scientific disciplines*' [6]. The strong programme, he says, 'possesses a certain kind of moral neutrality, namely the same kind as we have learned to associate with all *other sciences*' [7] and denial of its tenets would be a betrayal '. . . of the approach of empirical science' [8]. 'If knowledge could not be applied in a thorough-going way to scientific knowledge it would mean that science could not scientifically know itself' [9]. 'In a very orthodox way I have said: only proceed as the other sciences proceed and all will be well' [10]. 'If we want an account of the nature of scientific knowledge, surely, we can do no better than adopt the scientific method itself' [11].

Larry Laudan, Bloor's chief critic, is quick to spot the apparent circularity involved here [12]. He points out that Bloor seems to be assuming the very answer he is setting out to discover: Bloor's characterization of, and justification for, his tenets puts the cart before the horse. But Laudan also deploys a version of 'what science is like' in his own attempt to discredit Bloor's claims. He thus argues that not all science is causal; that impartiality occurs in no science of which he is aware; and that the reflexivity tenet is redundant if the objective is a generally applicable theory [13]. Laudan is especially troubled by the symmetry postulate because it runs against *what he regards as the best established precedents in the natural sciences* namely, that scientists invoke different causal processes to explain different phenomena. It would be absurd, suggests Laudan, to try and explain both gravitational and electrical phenomena using the same kind of cause. Of course, it is possible that Laudan has misread Bloor on this point. Bloor does not insist on the same cause for different phenomena, but on the same *kind of* cause for both true and false instances of the same phenomenon. A better analogy is not with the unicausal explanation of gravitational and electrical phenomenon, but with social interests giving rise to both N-rays and X-rays. Specifically different social interests might have operated in each instance, but social interests would nonetheless account for an instance of (perceivedly) true scientific knowledge as well as a false one. Bloor wishes to avoid the kind of situation where social interests are used to account for N-rays but rationality and logic are used to account for the emergence of X-rays.

Even at the heart of disagreements about how to find out about science, we find the protagonists utilizing versions of what science is (actually) like. Taken together, their arguments display the kind of variation in preconceptions about science which we found in Chapter 1. It is as if the protagonists are locked into a discourse which *forces* their use and invocation of one or another preconception about science. This observation further supports the argument that 'science' is best treated, not as a discoverable entity, but as a discursive resource. More importantly, the fact that the arguments of even these analysts have recourse to unexplicated notions of science raises the possibility that we are dealing with a concept which is deeply implicated in practices of argument.

RULES AND LOGIC

The call for a sociology of scientific knowledge attracted a lot of attention, not just because it proposed the sociological analysis of previously philosophical matters — the content and nature of scientific knowledge — but more significantly, because it emphasized the relativity of scientific truth. The implications were that scientific knowledge could no longer be assumed to be straightforwardly 'rational', that the application of 'reason' was no longer any guarantee of 'truth' and so on. In fact, this kind of relativity was no more than a particular case of a more widespread intellectual movement. In particular, SSK shows a marked affinity with a key notion in post-Wittgensteinian thought: scepticism about the view that practice (action, behaviour) can be understood in terms of following rules (guidelines, principles). In order to elaborate this point let us consider the rudiments of the sociology of scientific knowledge position on the nature of rules and logic.

In a famous parable by Lewis Carroll (used subsequently by Winch), Achilles and the tortoise are discussing three propositions — A, B and Z — which are so related, according to Achilles, that Z is claimed to 'follow logically' from A and B [14]. The tortoise agrees to accept A and B as true but asks what would persuade him to accept Z if he does not yet accept the hypothetical proposition C: If A and B are true, Z must be true. Achilles begins by asking the tortoise to accept C, which he does. He then says to the tortoise: 'If you accept A, B and C, you must accept Z'. When the tortoise asks why he must, Achilles says 'it is because it follows logically from them. If A, B and C are true, Z must be true. You don't dispute that I imagine?' The tortoise agrees to accept this last proposition and to call it D.

'Now that you accept A and B and C and D, of *course* you accept Z.

'Do I?' said the tortoise innocently. 'Let's make that quite clear. I accept A and B and C and D. Suppose I still refuse to accept Z.'
'Then logic would take you by the throat and *force* you to do it'

Achilles triumphantly replied. 'Logic would tell you "You can't help yourself. Now that you have accepted A and B and C and D, you must accept Z". So you've no choice you see.'

'Whatever logic is good enough to tell me is worth *writing down*', said the tortoise. So enter it in your book, please. We will call it E (If A and B and C and D are true, Z must be true). Until I have granted *that*, of course, I needn't grant Z. So it's quite a necessary step, you see?'

'I see,' said Achilles; and there was a touch of sadness in his tone. [15]

In Lewis Carroll's version, the story ends some months later when the narrator returns to find the couple still discussing the point, with the notebook nearly full. (A more realistic version would depict Achilles as altogether less patient: the story would end in the culmination of Achilles' frustration and his dismissal of the tortoise, certainly no later than about proposition G.)

The moral of the tale is that in principle there is nothing in the logic itself to guarantee the acceptance of a proposition or position. Rules and reasons do not themselves determine the position adopted by parties to an argument. This is so, in particular, because any justification of a particular logical connection is itself susceptible to justification. The search for final or absolute justification is therefore endless in principle. (The methodological horror which applies here is inconcludability — see Chapter 2). In practice, however, participants say that enough is enough, thereby appealing to one another's sense of 'what we could all reasonably expect to be the case'. Logic compels, it has been said, by the sanctions of our fellow men.

This view of logic supplants the notion of reason as a determinant of action [16]. Reason and logic are in principle insufficient to compel a particular course of action. (Action, like knowledge, is underdetermined by rules, logic and reason (observations).) Instead, logic and reason are key features of a discourse which is used to evaluate and characterize action. Action comes first, logic second, although this is not just a matter of temporal sequence. Rather we are talking of logic as an antecedent in the fullest sense. Its invocation as an antecedent is inevitably *post hoc* in that actions are envisaged and their grounds only subsequently grafted onto them. Now, of course, I may decide upon a course of action in virtue of my contemplation about 'whether or not it is reasonable'; I could say that I worked out 'what it would be logical to do'; that it 'made sense' to follow one path rather than another. But in all these cases, the envisaged action and the consequent actions are also antecedent to subsequent rationalization. In such cases, the logical schema are imposed upon actions which are imagined (in the light of past experience or whatever). The logic of the situation does not exist outside of descriptions and assessments of the action itself. Logic can not simply 'give rise to' actions.

One classic form of logical reasoning is the syllogism, an example of which is:

(1) All politicians are liars
(2) Mrs Thatcher is a politician
(3) Mrs Thatcher is a liar

The syllogism is such that the conclusion (statement (3)) is said to follow from two premises (statements (1) and (2)). In a manner analogous to Boolean algebra, the first statement defines a category (liars) within which a subcategory (politicians) is contained; the second statement identifies a member of this subcategory, which is also a member of the larger initial category. The difficulty is not that the reasoning is 'wrong', but that the implied necessity of logical deduction is superfluous [17]. We do not need to follow the logical steps of the syllogism to reach the conclusion, for we already 'know' that Mrs Thatcher is a liar *as part of* knowing that all politicians are liars. The deductive form of reasoning required by the syllogism is unnecessary for such knowledge. Its status appears to be a *post hoc* formalization (and hence justification) of something we were prepared to act upon anyway. Once again, we see that logic is subsequent to, rather than an antecedent of, the practical business of knowing something.

But supposing the example was modified. Suppose that although fully conversant with statement (1), we knew nothing about a Mr Bloggs. It was *only subsequently* revealed to us that Mr Bloggs is a politician. Would we not then say that the syllogism enabled us to conclude something we had not previously known, namely that Mr Bloggs is a liar? Surely, it might be claimed, logic helps us deduce something in this situation.

The example is important because it reveals a common confusion between logical and temporal links. In order to make the syllogism work for Mr Bloggs, we have in effect to extend the application of the first statement. For the generality (truth) of the first statement is contingent upon any further instances similarily fitting the mould. Not knowing anything about Mr Bloggs in advance, we might consider he was either a politician who lied or a politician who did not lie. The latter case would invalidate the initial premise of the syllogism. The former case would again make its 'application' trivial since we would not need the syllogism to tell us the terrible truth about Mr Bloggs.

Bloor also relates the argument that legal decisions and judgements can not and should not rest upon logical deduction from rules [18]. This is the way to real trouble. Instead, the decision should be made, and perhaps only later should the justification be retrospectively constructed. A judge in a recent public inquiry is reported to have similarly removed himself from the field of logical contention. Faced with the directly opposing claims of eminent scientists about the environmental impact of a nuclear power development, he sidestepped the truth of the matter: 'I may be right and I may be wrong, but I am never uncertain — I hereby find for the claimant'. The practical resolution of apparent deadlock was achieved by redefining the relevant decision criteria. Logic and truth were left behind in favour of considerations of decisiveness.

We have thus far exploded two myths about logic: the tortoise showed us that logic does not determine a particular course of action (practice, deduction, knowledge); our look at the syllogism suggests that logic is superfluous to a practical course of action. Together these lines of argument reinforce the notion that logic and reasoning have a function quite different from that normally attributed to them. Far from compelling particular courses of action, they form the *post hoc* rationalization for ordered practices and conventional ways of proceeding. Forms of logic, rationality and reason are then formal statements which reflect our acceptance of institutionalized practices and procedures. They are the vocabulary through and within which we reassert the primacy of consensual practice and institution.

REFLEXIVITY AND FEEDBACKING

The last tenet of the strong programme suggests that a form of reflexivity is necessary because otherwise sociology would be a standing refutation of its own theories. In other words, the supposition that sociology is immune from sociological analysis would be to suggest that it has achieved a form which places it above the knowledge enterprises it seeks to explain. It would thus constitute a special case, the very thing outlawed by the quest for a thorough-going sociology of knowledge. To make an exception of the sociology of scientific knowledge would be to curtail the general validity of its argument and we would then be back in the situation referred to (in Chapter 1) as Mannheim's Mistake: the explicit exemption of a particular kind of knowledge (in his case, mathematics and natural sciences) from the purview of sociological analysis.

An interesting question arises in respect of the possible outcome of this debate. What will mark the end of the Bloor–Laudan debate? A definitive answer would of course provide a test of their respective theories. The rationalists' answer seems straightforward: the inherent logic and rationality of their argument will ensure the correctness of their position. Further, the error of the sociologists is easily explained by their inattention to matters of logic and rationality, an oversight brought on, perhaps, by their (illegitimate) desire to extend the empire of their professional expertise. The strong programmers' answer is less clear-cut. For, to be consistent, they would need to admit that factors other than the intrinsic merit of their case will bring about a resolution [19]. They would grant that the competing interests of rationalists and themselves are at play, but would not easily be able to predict an outcome.

The inability of the strong programmers to predict the outcome is consistent with sociological scepticism about the idea of a 'definitive outcome'. We might all agree that a time will (has?) come when talk about and concern over the debate has ceased. At that point, the same mechanisms for rewriting the history of this episode will come into play as operate in the history of natural science. It will be possible, in principle, to portray the outcome as consistent with a victory for either the strong programme *or* the rationalist position. Whether Bloor or Laudan is correct is not some matter inherent to the argument, just awaiting the discovery of the necessary hidden manuscript. It is instead a matter of public perception in the light of concurrent and competing views and positions. Truth or falsity is perceived (and achieved) rather than inherent.

What then is the status of the strong programme in relation to sociological practice? Each of the four tenets which make up the strong programme take the form of a methodological injunction for the sociologist: the sociologist *should* be impartial and so on. But what is the relationship between such pronouncements and practice. If we carry through the scepticism about logic and reasoning to our consideration of rules, we see that these kinds of injunction do not *guide* practice, but merely provide *post hoc* justifications for the conventionally binding character of certain forms of practice. In what sense can we then suppose that the enunciation and elaboration of these tenets will generate the kind of SSK which supporters of the strong programme favour? Rules do not determine social action. Why then should these tenets lead to a certain kind of sociological research? The impressive analysis of logic produced by writers like Bloor suggests we understand these tenets as *post hoc* justifications of sociological research. They constitute, in other words, a resource for the characterization and evaluation of research practice.

We noted in Chapter 1 that the normative ethos of science was unsatisfactory in so far as it gave no good account of the generation of scientific knowledge. Indeed, it has been argued that the deliberate *transgression* of norms has lead to the generation of certifiable, valued new scientific knowledge [12]. We also noted that the philosophical quest for decision rules ran into trouble once it was recognized that 'true' knowledge could result from the deliberate disregard for what was perceived as the rational course of action. The policy implications of this line of argument are intriguing to say the least Feyerabend's (in)famous recommendation for science is 'anything goes' — that the specification of rules for rational procedure is counter-productive [21]. Does the same apply to the strong programme? Our 'reflexive' consideration of the status of Bloor's methodological injunctions suggests a further set of 'anti-guidelines' for conduct. Can we conclude that the health of SSK depends upon our deliberately contravening its tenets?

CONCLUSION

A central achievement of the sociology of scientific knowledge is its scepticism about the role of logic and reason especially in mathematics and science. This is elided with and derived from the scepticism about rule following of the later Wittgenstein.

SSK has established that the esoteric details of scientific activity (the process whereby knowledge about the world is produced, the work of connecting the left- and right-hand sides of our representation couple) is an appropriate focus of sociological interest. In particular, this chapter has outlined the argument for a first key inversion with respect to science. By looking at reason and logic, we find that reason, logic and rules are *post hoc* rationalizations of scientific and mathematical practices, not their determining force. Logic does not give rise to a particular deduction or proof but instead justifies the conventionally accepted manouevres which count as that proof. The implication for social science and more generally for all attempts to explain and account for phenomena — whether human, animal, mechanistic or inanimate, etc. (see Chapter 7) — is that we are not governed by logic, nor by rules nor reasons.

But at this point a crucial set of alternatives opens up. We could either abandon the attempt to explain science by logic (rules and reasons) in favour of some other expanans *or* we could abandon the attempt to explain science in this way at all. Advocates of the strong programme seem to come close to suggesting that we understand scientific activity (or, at least, the practical operation of logic) in terms of conventions. The important point is that pursuit of the strong programme means we remain committed to a particularly scientific notion — explaining — in trying to make sense of science. It is not hard to see the similarity between the explanatory format of Mertonian and strong programme accounts. 'Social interests' take the place of 'social norms', but otherwise the form of explanation is essentially unchanged. Is it wise to persist in this explanatory mould, or should we use the occasion of increased scepticism to explore some (more) radical alternatives to explanation altogether? As a first step in the search for alternatives to explanation, we begin, in the next chapter, to apply inversion and feedbacking to other aspects of the traditional idea of science.

NOTES

[1] R. K. Merton, 'Science and Technology in a Democratic Order', *Journal of Legal and Political Sociology*, **1**, (1942), p. 116. Article subsequently published as 'Science and Democratic Social Structure' in R. K. Merton, *Social Theory and Social Structure*, and as 'The Normative Structure of Science' in R. K. Merton, *The Sociology of Science: Theoretical and Empirical Investigations*, University of Chicago Press, Chicago (1973), pp. 267–278.

[2] T. S. Kuhn, *The Structure of Scientific Revolutions*, 2nd edition, University of Chicago Press, Chicago (1970).

[3] D. Bloor, *Knowledge and Social Imagery*, Routledge & Kegan Paul, London (1976).

[4] Ibid., Chapter 1.

[5] Ibid., p. 45.

[6] Ibid., p. 4.

[7] Ibid., p. 10.

[8] Ibid., p. 10.

[9] Ibid., p. 40.

[10] Ibid., p. 141.

[11] Ibid., p. ix.

[12] L. Laudan, 'The Pseudo-Science of Science?', *Philosophy of the Social Sciences*, **11**, (1981), pp. 173–198.

[13] Ibid.

[14] L. Carroll, 'What the Tortoise Said to Achilles' in L. Carroll, *Complete Works*, Nonesuch Press, cited in P. Winch, *The Idea of a Social Science*, Routledge & Kegan Paul, London (1958), pp. 55 ff. For a related demonstration of the conventional base for rule following, see the example (adapted from Wittgenstein and Winch) in H. M. Collins, *Changing Order: Replication and Induction in Scientific Practice*, Sage, London (1985), pp. 12–16.

[15] Winch, op. cit (note 14), p.56.

[16] David Bloor shows how this argument applies as much to mathematical logic as to the logic supporting the Azande oracle. Bloor, op. cit. (note 3), Chapters 6 and 7.

[17] Ibid., p. 117 ff.

[18] Ibid., p. 118.

[19] Actually, sophisticated strong programmers would object to this formulation since it implies that 'intrinsic merit' is separable/distinct from 'other factors'. They may wish to make the point that such 'factors' *constitute* value, truth, merit and the rest.

[20] M. J. Mulkay, *The Social Process of Innovation*, Macmillan, London (1972).

[21] P. K. Feyerabend, *Against Method,* New Left Books, London (1975).

4

Inverting Nature:
Discovery and Facts

Logic and reason are just one focus of the challenge to the 'received view' of science [1]. In the last chapter, we looked in particular at the way the strong programme in SSK took issue with the view that scientific knowledge was generated as a result of rational (reasonable, logical) extrapolation from either (or both) existing knowledge or observations of the world. In terms of our representation couple, critical attention was directed towards the *character* of the link between right-hand side and left-hand side:

knowledge ——————— the world
new knowledge ———— old knowledge

The strong programme argued that the creation of these links could not be assumed to be the upshot of a rational process, if that meant their exclusion from sociological investigation. It was argued that the sociological perspective enabled us to understand rules, reason and logic as social conventions for accomplishing and creating these links.

We thus see that sociologists have unequivocally rejected the assumption in 'the received view' that the production of knowledge about the world, that is, the establishment of connections between right- and left-hand sides of our couple, is not amenable to sociological study. However, many seem uncertain about taking issue with a further key assumption, that the world exists independently of, and prior to, knowledge produced about it. Although they are clear about the need for a sociological position on the way the link is forged, they are less clear about how there comes to be a right-hand side in the first place.

This ambivalence in sociologists' accounts is most evident in their programmatic statements, the introductory sections and conclusions to empirical studies. For example:

> Occasionally, existing work leaves the feeling that reality has nothing to do with what is socially constructed or negotiated to count as natural knowledge, but we may safely assume that this impression is an accidental by-product of over-enthusiatic socio-logical analysis, and that sociologists as a whole would acknow-ledge that the world in some way constrains what it is believed to be. [2]

> there is ... nothing in the physical world which uniqely deter-mines the conclusions of (the scientific) community. It is of course self-evident that the external world exerts constraints on the conclusions of science. [3]

> empirical facts *by themselves* do not determine knowledge claims. [4]

> the natural world has a small or non-existent role in the construc-tion of scientific knowledge. [5]

These statements suggest that while traditional perceptions of the relation-ship of knowledge to the natural world have been considerably weakened, there remains uncertainty about the consequent status of 'the natural world'. Despite the apparent radicalism of its stance on the relativity of scientific truth, this ambivalence raises the possibility that SSK has done little to revise basic ontological commitments. Indeed, recent work in SSK has been dubbed 'epistemologically relativist and ontologically realist' [6]. This is curious given that a major thrust of post-modern critiques of science is to suggest the essential equivalence of ontology and epistemology: how we know *is* what exists.

In order to begin to extend the radical potential of sociological studies of scientific knowledge, this chapter applies the principle of inversion to a slightly different aspect of the representation couple, namely the *direction* of the connecting link. Whereas we tend to think of left-hand side entities as arising from pre-existing right-hand side entities,

representation ⟵——————— object

scientific knowledge ⟵——————— the natural world

some parts of work in SSK suggest the direction of the arrow be reversed, that right-hand side entities are constituted (constructed, defined, accomplished) in virtue of the left-hand side.

scientific knowledge ——————⟶ the natural world

We examine the sense and consequences of the arrow's reversal by considering the notion of discovery.

DISCOVERY

Discovery is central to common conceptions of science. New discoveries are generally reckoned to be about objects, events and processes of which we have no previous knowledge. And science is the social arena where discoveries are thought most likely to occur, since science is thought to embody the most reliable and efficient procedures for the generation of new knowledge. This image is reinforced and perpetuated by popular accounts of science in various media. (The BBC television programme *Tomorrow's World* is perhaps archetypal.) The media are about news and new discoveries are quintessentially newsworthy. Discoveries of the kind associated with science do not happen every day.

The metaphor of scientific discovery, the idea of dis-covering, is precisely that of uncovering and revealing something which had been there all along. One removes the covers and thereby exposes the thing for what it is; one pulls back the curtains on the facts. The image derives in part from the notion of geographical discovery. One travels to a distant place and finds (comes upon or otherwise stumbles over) what was already there. The crucial part is the prior existence of the discovered object. The central assumption in discovery is that the discovered object is antecedent, that it enjoyed an existence before travellers came across it. The rhetoric of this ontology portrays the objects of discovery as fixed, but the agents of discovery as merely transient. A common analogy is the idea of a scientist sailing upon an ocean, from time to time coming across islands of truth.

What constitutes a discovery? I might say that I have just recently discovered Richard Strauss's *Four Last Songs,* or that the cost of living in Kenya is half that of England, or a hole in my sock. Obviously, it is unlikely that any of these actions will be accorded the status of scientific discovery. Why not? The first two examples might not be regarded as discoveries because they are things and facts which are already well known (and known about) by others. Hence, whether or not some observation or state of knowing amounts to a discovery depends importantly upon *for whom* it is a discovery. The point is nicely captured in the old joke about the father who complains: 'My teenage son thinks he has just discovered sex, but I know I discovered it twenty five years ago'. By contrast, the observation of the hole in my sock might be entirely unknown to any other living sole (ha-ha). But this observation is unlikely to be of much interest to anyone else. Hence, an obvious further requirement for an observation to acquire the status of scientific discovery, is that it has a particular degree of significance for those to whom it is news. We see, then, that observations must be perceived as both novel and significant, before they can be accorded the status of discovery.

The sociologist wishes to know what counts as 'novel' and 'significant'. Under what circumstances and in what ways do certain definitions of novelty and significance hold sway, perhaps at the expense of those of rival scientists? What is the process whereby a discovery claim becomes accredited? What kinds of resource must be mobilized and what sort of discovery accounts advanced in order to establish the presence of a new and significant phenomenon? Importantly, the answers to questions about novelty and significance are bound up with answers to the prior question: how does there come to be a phenomenon 'out there' in the first place?

Representation constitutes object

The idea of reversing the arrow is to suggest that objects are constituted in virtue of representation. As a preliminary (negative) argument in favour of this way of looking at things, let us consider the deficiencies of the alternative, objectivist position.

An old philosophical chestnut asks whether or not a tree falling in a far away forest makes a noise: in what sense is it sensible to say that the tree makes a noise when nobody is there to hear it? The more general question, of course, is: in what sense can we say the tree exists even though nobody is there to observe it? Clearly, our absence from the scene does not prevent us from conjecturing the existence of an object. Our ability to speak as if realities exist independent of our knowing them is a key function of language and representation. But can an object exist independent of our practices of representation?

In a classroom situation, the challenge to students (those with realist proclivities) is to demonstrate the existence of any object (fact, event, thing) without recourse to a form of representation. I point out that a successful demonstration would stand as a strong argument that facts can indeed have a life of their own. And it would follow that we would have to concede that facts (objects, things) can be antecedents; that representative practices (i.e. the social context) are merely adjuncts, passive tools for revealing the character of these things for what they actually are.

Students quickly see the difficulty of the task. Indeed, no one has thus far succeeded in demonstrating the antecedent existence of a fact or a thing independent of some representative practice. Students are often less willing to grant the *impossibility* of the task. They reassert, sometimes with much vehemence, their belief in the prior, independent, objective existence of objects in the world. They argue, in effect, that the problem I am raising is a merely technical difficulty: that their (temporary) inability to produce an adequate demonstration is insufficient refutation of the objectivist position (see Chapter 2: managing the methodological horrors). I point out that this is mere assertion, rather than demonstration, and that in any case they employ the rhetoric of representation when describing the (supposedly) objective state of affairs to which they are committed. This sometimes leads to interesting attempts to meet the challenge through actions which they feel might not be construed as representation. For example, gestures will be used to point to objects in a bid to establish their self-evident character, as if gestures were any less a manner of representation than spoken or written language. Or again, in the same spirit of exasperation experienced by Achilles in Chapter 3, students suggest that a punch in the face might finally convince sceptics of the reality of the physical world. In this unhappy event, the bruised relativist would again have to remind his audience that physical assaults also constitute a form of representation/communication.

If we grant that this whole exercise fails to refute objectivism, we should note that it equally pinpoints the massive difficulty in establishing objectivism. On this basis we are at least entitled to entertain the alternative (constructivist) position as a heuristic. If we grant the possibility that representational practice actively constitutes the world, how are we to understand the ways in which discovery claims make out aspects of the world as novel and significant?

A simple initial response to this question is that what counts as novel and significant will depend upon the social context in which these claims are made. It is important to note, however, that social context can not be simply construed as a mere appendage to the fact of a discovery, since that would imply the possibility of making observations in isolation from social context. This latter notion of social context is consistent with the argument that meanings can be derived through knowledge of context. The formula sometimes recommended is 'language (representation) + context = meaning'. Of course, asserting the relevance of 'context' is an important way of reminding ourselves of the indeterminancy of language. But when 'context' becomes a principled solution for the methodological horror of indexicality

(Chapter 2), we are in danger of overlooking the way in which context and meaning are deeply implicated in one another. In our examples of discovery, therefore, we shall explore in detail the sense in which social context *constitutes* the discovered object.

The construction of 'America'

Brannigan's discussion of Columbus and the 'discovery' of America shows the important and profound sense in which discoveries occur within a social context [7]. In the case of Columbus, the social context comprises the several decades of preparation, the organization of expectations and activities during the voyage and the complex work of recording, publicizing the events of the voyage in its aftermath. For almost twenty years, Columbus had failed to generate support for the venture to reach a mass of land assumed to be the east coast of the Indies. He had been turned down by private sources, by the Portuguese crown and by the Spanish crown, the latter only relenting after a long period of deliberation and threats that possible fruits of the mission might go elsewhere.

The voyage was planned drawing upon particular assumptions about geography, the whereabouts of the land mass to be discovered and the likelihood of meeting natives. The voyagers carried trinkets, beads and buttons and budgeted for provisions for a round trip of 4000 miles. The voyage itself was organized in terms of the participants' awareness of the nature of the mission, their expectations, the incentives for sightings of land and so on. Finally, after leaving the discovered lands and setting sail for home, there continued an elaborate process of manoeuvring for institutional recognition of the discovery. This is crucial, since as we have already seen in the simple examples above, the discovery would not count as such without institutional approval of its claimed significance.

On the return journey, Columbus wrote messages and dropped them overboard in sealed caskets. The messages urged the finders to communicate the news to the Spanish authorities, and mentioned the likely financial reward for so doing. Against their fears, the putative discoverers (for such they were at this point) reached home safely. Columbus then set in train a long, complicated sequence of publication and communication of the news of the discovery. His initial letter to the Spanish court was widely reprinted throughout Europe, culminating in the Vatican's approval of Spanish claims to the discovered lands.

Two crucial points emerge from this example. Firstly, discovery is a process rather than a point occurrence in time. It is a process of planning, anticipation, soliciting support and obtaining institutional approval of a claim or definition (that a discovery has occurred). When we say that

Columbus discovered America, we are summarizing the outcome of an extended period of claims and definitions, and we are opting for a particular outcome sanctioned at a particular time by a particular social agency. Secondly, the discovery process extends in time both before and after the initial announcement or claim. We have already noted that the discovery was rooted in decades of preparation and quest for support. But the process of moulding and shaping the nature of Columbus's achievement continued long after the Vatican's seal of approval. Columbus made further voyages, persisting with the assumption that the lands he had found were islands off the east coast of The Indies. Almost ten years after Columbus's first voyage, another explorer, Amerigo Vespucci reported that he had found an extended land mass, thus countering Columbus's assumption. This claim met resistance, Brannigan suggests, because the notion of finding new parts of the world contradicted the dominant Christian notion that the earth was a uniform, known entity. Eventually, however, it was Vespucci's version which held sway. The histories of the sixteenth century rewrote Columbus's achievement. He had, it was now said, discovered America — a hitherto unsuspected land mass — contrary to his own specific claims.

Brannigan notes that the whole episode is just one of several related stories about the upshot of voyages to 'land masses across the Atlantic': the neolithic Siberians are associated with the idea of an Aleutian land bridge; pre-Christian Phoenicians and their tales of a great river (the St Lawrence) and a new continent; the Irish monks and a new Iceland; Norsemen, led by Leif the Lucky, in discovering a Greenland, Vinland and so on [8]. The point is that although there is a sense in which each of these travellers could have claimed to have discovered America, neither they, nor spokespersons subsequently acting on their behalf, could or did pursue a social process culminating in the public legitimation of their claims. Deprived of the resources available to Columbus, and deprived, in particular, of the social organization of beliefs which subsequently shaped and reshaped the Columbus story, these alternative claims failed.

The strength of the successful account — that Columbus discovered America — is its entrenchment. The stability of this particular factual claim is precisely a reflection of the enormous amount of work which is now required to deconstruct it. As champions of pre-Columbian discovery accounts know well, they have a substantial fight on their hands. They have to challenge widespread popular beliefs, the certified histories and records of most European countries, and a vast established network of institutionalized celebration of the image of Columbus (Columbia University, Columbia pictures, etc., etc.). As a kind of shorthand for the massive work required to undermine all this, we refer to the discovered object as a fact. We say it is true, a thing (Latin: *res*) which *res*ists our efforts to overcome or deconstruct it [9]. The degree of resistance is a direct reflection of the work invested during a long, complex social process of definition.

It is important to distinguish between the fact of the discovery (that it was due to Columbus, in the current consensus) and the fact of the discovered object. The former is the weaker argument that several competing versions are all claims for the same discovered reality. This seems to be the implication of the argument that it was Vespucci rather than Columbus (or vice versa) who made the discovery. The problem with this way of looking at it is that it presumes the character of the discovered object, i.e. America is the same for Vespucci as for Columbus. But, as Brannigan's discussion of this and other discovery claims shows in detail, the character of the discovered object is the *upshot* of the process of defining and articulating what that object is. Vespucci's object was fundamentally different from Columbus's object [10]. Only in virtue of our accepting the fact of the discovered object, can we contrast and compare different claims 'about it'.

This supports our earlier suggestion that the study of how objects come to be constructed 'out there' should not be neglected in favour of attention to the process of defining novelty and significance. The fundamental and prior task is to understand the process culminating in the fact of the discovered object, rather than the fact of the discovery, since the latter tends to assume the prior existence and character of the object.

The more prosaic implications are clear: familiar questions about *who* made a discovery, or *when* it was made, illegitimately imply the antecedent existence of a constant object awaiting its revelation. As we have seen, the fact of an object is the temporarily stable upshot of a complex social process. Moreover, this process continues long after an initial discovery claim; 'the object' has and will continue to change.

The example of Columbus and the 'discovery' of America suggests that in our efforts to understand the social basis of discovery we cannot presume the character of a discovered object; the fact of a discovered object and its character is the achievement of the discoverer's (and/or others') claims and definitional work. To return to the representation couple once more, this is equivalent to saying that we can not presume the existence or character of the right-hand side. It is not that the (objective) right-hand side pre-exists our (human) efforts to come across it. Rather the right-hand side is the end result or accomplishment of work done by participants. Baldly speaking, discoverers create, rather than merely produce accounts of, the right-hand side.

In the next chapter we shall see that the same kind of inversion can be usefully applied to a much wider class of argument, explanation, understanding, perception and so on. For now, however, our concern is how this inversion is managed and sustained by discoverers. How, in other words, do discoverers make claims for the realist (received) view of objects which, our discussion suggests they have artfully created. To tackle this issue, we consider the discovery of pulsars: this moves us from tales about journeys to a distant shore, to tales about journeys within a network of astrophysicists and radio-astronomers.

A STORY OF PULSARS

On the 24 February 1968, *Nature* carried an article co-authored by Hewish, Bell and three other members of the Cambridge radio astronomy group [11]. They claimed to have discovered unusual rapidly pulsating radio sources (only subsequently referred to as pulsars).

The straightforward announcement in *Nature* conceals an extra-ordinary complexity in the accounts and recollections of participants. It is quickly evident, for example, by examining in detail both written and verbal accounts of the discovery, that there are clear discrepancies in participants' recollections of events leading up to the discovery. These discrepancies were not eased by participants' subsequent attempts to recover what actually happened. Nor is this just an academic concern about historical accuracy. This particular situation was further confused when the award to Hewish of the Nobel Prize for physics (jointly with Martin Ryle) in 1975, resurrected the bitterness which had laid dormant since the initial announcement. There then followed accusation and counter-accusation about the real circumstances and sequence of events leading to the discovery [12].

The importance of controversy about discovery within the community of practitioners is its objectivity-reinforcing function. To argue about *who* discovered pulsars and *when* this discovery was made, has the function of reinforcing the antecedent objective status of the discovered objects — pulsars — themselves. A controversy of this kind involves participants in the pursuit of precisely those questions which it was recommended (above) that we (*qua* analysts) try to avoid.

Of course, it is impossible to proceed without some kind of account. Consequently, I begin with what can be considered a 'working account', the status of which is ultimately questionable, but which serves to introduce or provide the background necessary for the ensuing analysis [13].

The period between the setting up of the recording equipment and the telescope — an 81.5 MHz antenna constructed in a rectangular array of 2048 dipoles covering an area of 4.5 acres — in July 1967, and the date of the *Nature* publication, can be divided into five more or less distinct phases. The *first phase* is marked by the initial recognition of an anomalous trace on routine chart recordings. But the anomaly was not at that stage regarded as being sufficiently worthy of special investigation. Unusual traces were not an uncommon feature of these experimental observations. There was no special discussion about the trace between Hewish and Bell and no reason to think the occurrence merited any special attention. During a *second phase,* the repeated occurrence of the trace prompted [14] Hewish and Bell to consider it worthy of investigation. However, this was not a particularly remarkable decision. Such investigations frequently revealed strange traces to be the result of interference. High speed recordings were planned but there was still no conviction that the strange signals were especially significant. Consequently, at this stage, discussion did not extend beyond Hewish and Bell, who were in any case meeting on a daily basis. *Phase three* followed the recognition of the pulsed nature of the radio emission shown up by the successful high speed recordings. Three researchers were persuaded to join Hewish and Bell in pursuing lines of investigation to elicit the origin of the signals. At this time, other members of the group gradually became aware that something unusual had been found and was being investigated. The three researchers were co-opted into the study with some urgency because the signals were then thought to be the result of some temporary flare-up (again a not uncommon phenomenon). But the unexpected character of the pulsed recording led to a sceptical view that the findings were likely to be spurious or the result merely of an unusual type of interference. Hence the desire at this stage to keep quiet about the investigations in case the results proved to be entirely trivial. Few members of the group had detailed knowledge of the investigations. The strongest factor preventing the spread of information outside the group at this stage was a tacit understanding among members that they should keep quiet about any new work that went on. One line of criticism subsequently directed towards the group was that publication should have come in the aftermath of the first instance of pulsed recordings. But the Cambridge group were wary of publishing observations which might have later proved trivial. *Phase four* refers to the period when the researchers had largely discarded the idea that the signals resulted from interference or some temporary outburst of radio activity. The most viable explanation at this stage was that the signals represented communications from another intelligent civilization: the chart recordings were annotated with the legends — 'LGM1', 'LGM2' — ironically denoting the possibility of 'Little Green Men'. The serious implications of this possibility meant that the findings could not be presented in the normal way. So the small core of researchers with detailed knowledge of the investigations felt it necessary to make a conscious decision to restrict all mention of their work. This now provided a constraint on the spread of information in addition to the tacit understanding among those with less detailed knowledge of what was going on. Once the possibility of intelligent signals was discounted, however, there was still no divulgence of information. This *fifth phase,* argued some critics, should surely have been the time to publish the findings. But justifications for reticence at this stage were primarily in terms of the desire to conduct further experiments on the newly discovered phenomenon before releasing news which would enable other better equipped observatories to scoop much of the credit for the initial investigation. In the face of much criticism, Cambridge participants maintained that the entitlement to exclusive follow-up investigations was their natural privilege.

The rhetoric of criticism and justification which surrounded this episode is especially revealing in terms of our earlier (Chapter 1) discussion of the explanation of scientific behaviour by social norms. Mertonian sociology of science suggested that scientific action was the result of various institutionalized imperatives to which members of the scientific community were beholden. It is not appropriate to assess here whether or not the various parties to the pulsar dispute actually conformed to or deviated from social norms. It is apparent, however, that descriptions and evaluations of behaviour were very much couched in terms of Merton's normative ethos. For example, Hewish and his group were said, in effect, to have violated the norm of communalism in refusing to share/divulge their findings at an earlier stage in the proceedings. At the same time, however, it was typical of the Cambridge group's response that they could defend their actions effectively in terms of another of the Mertonian norms, that of organized scepticism. It was necessary, they argued, to be absolutely certain that they had indeed discovered a new astrophysical source, on pain of misleading the entire scientific community and possibly wasting scientific resources in a wild goose chase.

Their flexible use and invocation by participants themselves suggests that social norms are far from the straightforward determinants of scientific action portrayed in the Mertonian account. Instead of causing/determining scientists' actions, it seems that these norms are part of an evaluative repertoire drawn upon by scientists in their characterization, description and evaluation of their own and others' behaviours [15]. Once again, we find in this example, further evidence to suggest that entities routinely and traditionally taken as causes of participants' action are more usefully understood as a resource drawn upon in participants' discourse.

The substance of criticisms directed to the Cambridge group hinged largely on what constituted 'the discovery'. Participants themselves variously reconstructed events leading up to the discovery in ways which seemed to them to produce a 'logical' sequence, but which was often quite different from other available accounts. This uncertainty about the sequence and date of the discovery provided the tangible focus for allegations of undue delay and secrecy. Even seven years after the announcement, a prominent critic was still claiming that 'information concerning the observation of November 28th (1967) is still vague. Hewish does not make the situation at all clear' [16].

Each of the five phases outlined above corresponds to the progressive involvement of an increasingly wide circle of core researchers, to different types of constraint on the communication of knowledge and to different aspects of criticisms of and justifications for these constraints. Most significantly, *each of these phases corresponds to a different kind of (putatively) discovered object.* Only in retrospect can we (do we) 'recognize' that all along the participants were on the track of the same object: a 'pulsar'. And this retrospective view is only possible in virtue of our agreement with the discovery claim (made in the 1968 *Nature* article). But we see that the character of the object, and whether or not there was an object, changed with changes in (what might loosely be called) the social context: that is, changes in the personnel involved, the equipment brought into play, the literature consulted, constraints on communicating knowledge, criticisms and justifications for behaviour. Before the very possibility of an 'it' had begun to stabilize, the object (and the non-object) enjoyed at least five separate incarnations:

(1) an unusual trace; a non-object
(2) possible interference
(3) a temporary flare-up; or unusual interference
(4) communications from another civilization (little green men)
(5) new kind of pulsating radio source

It is worth noting that even without the assumption of hindsight — that it was after all a pulsar — the delineation of these phases is still not entirely free from the logic of retrospection. For example, to delineate phase (1) and to describe the object in that phase as an unusual trace, is to impute more definitiveness to the sequence than was (perhaps) current at the time. It is also worth stressing that this five-phase division covers only a short period immediately before the *Nature* announcement in 1968. By analogy with Columbus, we must recognize that determinations of social context (the organization of the Cambridge research team, its mores with regard to communication of knowledge to competitors, the equipment made available, the state of the art *vis-a-vis* recording apparatus, the expectations and beliefs of the principal players) have their origins in the pre-history of radio astronomy and astrophysics. Similarly, the definition of the object has enjoyed a chequered career since the discovery announcement. It has been a white dwarf star, a rotating neutron star, a neutron star with a satellite, the plasmic interaction between binary neutron stars and so on. For the record, the current consensus at this level of definition favours a rotating neutron star. Should this temporarily stable construction be overthrown (revised) in the future, the usurpers will have to deconstruct nearly twenty years worth of mobilization of resources and argument.

CONCLUSION

The main conclusion from our examples of discovery is that the existence and character of a discovered object is a different animal according to the constituency of different social networks. And by social network we refer to the beliefs, knowledge, expectations, the array of arguments and resources, equipment, allies and supporters, in short, to the whole local culture, as much as to the identities of individual participants. Crucially, this variation undermines the standard presumption about the existence of the object prior to its discovery. The argument is not just that social networks mediate between the object and observational work done by participants. Rather, the social network constitutes the object (or lack of it). The implication for our main argument is the inversion of the presumed relationship between representation and object; the representation gives rise to the object:

$$\text{representation} \longrightarrow \text{object}$$

Of course, this is a profoundly unsettling way of looking at things. Our 'instinctive' reaction — 'Surely this is not right' — precisely mirrors our commitment to, and familiarity with, the unidirectional link in the other direction. We somehow feel that representation can only follow from objects, not the other way around. Although it is small comfort to say so, such profound discomfort is a good index of the extent to which we take for granted the view which is being inverted.

The more immediate difficulty is this. If we claim that discoverers (or, more exactly, the social networks within which discoverers operate) actively constitute their object, how then do they manage to convince themselves of the reverse? In the face of our perfectly reasonable articulation of this particular methodological horror, what sustains and continually re-accomplishes the prior existence of discovered objects? In the next chapter we tackle this issue by considering some general features of the structure of scientific discourse.

NOTES

[1] We shall consider, in Chapter 7, the argument that logic and reason is a particularly strategic focus for the social study of scientific knowledge.

[2] B. Barnes, *Scientific Knowledge and Sociological Theory*, Routledge & Kegan Paul, London (1974), p. 7.

[3] M. J. Mulkay, *Science and the Sociology of Knowledge*, Allen & Unwin, London (1979), p. 61.

[4] B. Harvey, 'Plausibility and the Evaluation of Knowledge: a case Study of Experimental Quantum Mechanics', *Social Studies of Science*, **11** (1981), p. 95.

[5] H. M. Collins, 'Stages in the Empirical Programme of Relativism', *Social Studies of Science*, **11** (1981), p. 3.

[6] A phrase due to Donald Campbell. See D. T. Campbell, 'Science's Social System of Validity-Enhancing Collective Belief Change and the Problems of the Social Sciences' in D. W. Fiske and R. A. Shweder (eds.), *Metatheory in Social Science: pluralisms and subjectivities*, University of Chicago Press, Chicago (1986), pp. 108–135.

[7] A. Brannigan, *The Social Basis of Scientific Discoveries*, Cambridge University Press, Cambridge (1981), Chapter 7.

[8] Ibid.

[9] Cf. B. Latour and S. Woolgar, *Laboratory Life: the construction of scientific facts*, 2nd edition, Princeton University Press, Princeton (1986).

[10] Brannigan, op. cit. (note 7).

[11] A. Hewish, S. J. Bell, J. D. H. Pilkington, P. F. Scott and R. A. Collins, 'Observation of a Rapidly Pulsating Source', *Nature*, **217** (1968), pp. 709–713.

[12] S. Woolgar, 'Writing an Intellectual History of Scientific Development: the use of discovery accounts', *Social Studies of Science*, **6** (1976), pp. 395–422. S. Woolgar, *The Emergence and Growth of Research Areas in Science with Special Reference to Research on Pulsars*, Unpublished Ph.D. thesis, University of Cambridge (1978).

[13] Woolgar (1976), op. cit. (note 12).

[14] As if a trace could prompt! See Chapter 5.

[15] M. J. Mulkay, 'Interpretation and the Use of Rules: the case of the norms of science' in T. Gieryn (ed.), *Science and Social Structure: a festschrift for Robert Merton*, Transactions of the New York Academy of Sciences Series 2, **39** (1980), pp. 111–125.

[16] F. Hoyle, private communication to the author, 30 April 1975.

5

Arguing Science: Discourse and Explanation

In the last chapter we proposed an inversion of the relationship between objects in the world and their representation. It was suggested that representational practices constituted the objects of the world, rather than being a reflection of (arising from) them. At the same time, we noted that the apparent absurdity of this (inverted) point of view was a good measure of our commitment to the traditional standpoint.

It should be emphasized that this 'commitment' neither stems from nor necessitates a conscious declaration of allegiance to a realist ontology. It is not that we have made a deliberate decision to opt for this way of looking at things. Rather, as Bloor's analysis of logic reminds us, our 'commitment' is a measure of the way we organise our practical actions; realist ontology is a *post hoc* justification of existing institutional arrangements [1]. We think that objects precede and give rise to their representation precisely because this is the way we happen to organize our perceptions of the world, our negotiations with one another as to how actions are to be conducted and sanctioned, our expectations and so on. Given the entrenchment of these institutional arrangements, it is no wonder that an alternative *post hoc* justification appears absurd.

This chapter discusses some of the ways in which our embroilment in (and dependence upon) 'institutional arrangements' makes the inversion suggested in the last chapter seem absurd. We look at the way in which science can be construed as a discourse in and through which the prior existence of objects (things) is accomplished. In the terms of Chapter 2, we survey some aspects of scientific discourse which function to keep the methodological horrors at bay.

SPLITTING AND INVERSION

One way in which discourse constitutes its objects is suggested by the splitting and inversion model of discovery [2]. Recalling from the last chapter that discovery is a process rather than a point occurrence, we can propose a five-stage model which can be represented schematically as follows:

(1) document
(2) document → object
(3) document   object
(4) document ← object
(5) 'deny (or forget about) stages 1–3'

In the first stage, the scientists have documents (traces); in the case of the discovery of pulsars these comprise the charts from the telescope recorder, but might also include other publications, papers, previous results, the telescopes themselves, other apparatus, what Hoyle says and so on. At stage (2), participants use (some of) these documents to project the existence of a particular object (in this case interference, or an astrophysical phenomenon or whatever). Importantly, the object is created and constituted out of the documents available to the researchers. At stage (3), splitting occurs. Although the object was initially constituted in virtue of the documents (and more generally of the social network of which they are a part), it is now perceived as a separate entity, distinct from those documents. The object now has a life of its own. Indeed it is just one short step from possessing an infinite history; it is about to acquire the status of antecedent! In stage (4), the relationship between documents and the object is inverted. Whereas the object was constituted on the basis of the documents in step (2), it now seems as if the object (which was there all along) had given rise to the documents! It is at this point that the documents most obviously appear to take on the character of 'representations' or 'traces'; they are no longer just documents, they become documents *of* something. Step (5) is crucial. In order to maintain the inverted relationship of step (4), it is important to play down or minimize all reports which draw attention to the earlier steps (1), (2) and (3). Step (5) thus comprises the minimization, denial or backgrounding of all component parts of the process. Step (5) rewrites history so as to give the discovered object its ontological foundation.

Again, this scheme is unlikely to sit comfortably with taken-for-granted perceptions about the relationship between observations and objects in the natural world. One common reaction is to deny steps (1)–(3). And this is precisely the function of step (5). Or we might say that such is our conviction about the correct order of the relationship between representations and the natural world, that steps (1)–(4) occur very quickly. We are used to moving from documents to the idea of the objects which gave rise to them without blinking an eye.

Although the splitting/inversion model was developed for understanding the process of discovery, it is easy to see its application to representation in general. In other words, it is a model not just of the constitution of a discovered object but of all attempts to establish the antecedence of objects and things, to render them fixed (and objective) for a whole variety of purposes. These attempts range from strategies of causal explanation to the practical character of perception and interpretation in general.

One important feature of splitting and inversion is the way in which the reversal of the connection between observation and object also entails the removal from the scheme of any constitutive activity on the part of the observer. Construing the prior existence of the object entails the portrayal of the observer as passive rather than active. We thus see the rhetorical importance of the antecedence of the object in the way it implicates a particular conception of the agent. Once the object is construed as pre-given, fixed and antecedent, the involvement of the agent of representation appears merely peripheral and transitory. It is as if observers merely stumble upon a pre-existing scene.

As this suggests, we can anticipate the role of the agent to be of considerable strategic importance in the discourse of science. The presence or absence of the agent, his/her/its (see below) involvement in the work of representation is intimately bound up with the ontological robustness of the object itself. For a first demonstration of this, we turn to a consideration of the use of modalizers in discourse.

MODALIZERS

We can think of any scientific statement as involving a claim about the relationship between two component elements: A and B. For example: pulsars *are* rotating neutron stars; TRF(H) *is* a peptide; copper zirconium *does not behave like* other metal alloys. (The claimed relationship is emphasized in each case.) The claim thus comprises

A . B

This claim may enjoy a particular factual status at any one time, ranging from conjecture to documented report to a generally accepted fact. However, the factual status can change by the addition or subtraction of *modalizers*:

M . (A . B)

For example, contrast

(1) 'The glass transition temperature for this sample is 394 degrees C.'

with

(2) '*Giessen claims that* the glass transition temperature for this sample is 394 degrees C.'

and:

(3) 'Giessen claims that the glass transition temperature for this sample is 394 degrees C *because he wants it to be consistent with his previous results.*'

or again:

(4) '*It's in Giessen's interests* to suggest that the glass transition temperature for this sample is 394 degrees C.'

The sense (factual status) of the example (1) is altered by the addition of 'Giessen claims that' (2). This is further changed by the addition of a modalizer such as 'because he wants. . .' (3), or in a way directly comparable with the attempts of some sociologists to explain scientists' actions, by the use of 'It's in Giessen's interests to suggest that. . .' (4). Note how a fairly cursory reading of these examples suggests that the effect of adding modalizers is progressively to denigrate or downgrade the factual status of the initial claim. The argument along these lines is that the statement (2) introduces the identity of the agent (Giessen) and his activity ('claims') into a statement, the facticity of which we might expect not to depend on agency; statements (3) and (4) draw attention to possible motives for the agent's activity.

However, some caution is necessary in merely equating addition of modality with downgrading of facticity. For example, it may be that Giessen's involvement provides a way of buttressing a previously unsupported claim: 'If anyone should know about this stuff it is Giessen and he claims. . .'. The point, then, is not that the addition of modalizers changes the factual status in a particular direction, but that the inclusion of modalizers provides the grounds for a reassessment of the previously assumed (or suggested) factual status.

This caution is necessary in view of the observation that minor adjustments to the statements themselves can have a range of effects. Consider the following examples:

(5) human beings have the capacity to classify phenomena into groups [3]
(6) human beings *are thought to* have the capacity to classify phenomena into groups
(7) human beings have the capacity to *'classify'* phenomena into groups

Statement (6) revises the nature of the relationship between the two main components of the initial statement (5). Statement (7) revises the nature of one of the components of the initial statement by the use of a classic modalizer: quotation marks.

The main focus for our observations about the operation of modalizers is the way they draw attention to the existence and role of an agent in the constitution of a fact or factual statement. In general, then, we could say that the facticity of a statement can be enhanced (or lowered) by the deletion (or inclusion) of:

(a) reference to *agency* (the discoverer, scientist, author)
(b) reference of the *agent's action* (claiming, writing, constructing, etc.)
(c) reference to *antecedent circumstances bearing upon the agent's action* (his/her motive for making the claim, the interests served by acting in this manner, etc.)

The last element is interesting because it highlights what are recognizable as scientists' own efforts at sociological analysis. Indeed, the whole structure of modalizing work suggested by these three elements amounts to no less than participants' own attempts to do folk social science, i.e. to fashion associations between actors, their actions and the circumstances which might account for these actions [4]. The inclusion of the three elements can be heard as a deconstruction of a purportedly objective claim, just as their deletion can be understood as an attempt to objectify (that is, to render as an object, to put beyond the influence of agency) a statement initially wrapped up in irrelevancies.

Finally, it must be noted (in anticipation of our mention of replication later in this chapter) that the factual status of statements can vary with no apparent addition of a modalizer. Contrast

That's a bloody pulsar

with

That's a bloody pulsar [5]

The irony of the latter contrasts markedly with the naive optimism of the former!

THE TRIANGULATION OBJECTION

The implication that the objective world is constituted in and through discourse runs counter to our common-sense views, based as they are upon realist epistemologies. One obvious objection to the constitutive argument is that the objective world is available (describable, reportable on) by means of a *variety* of reporting and recording strategies, and by a *variety* of representations; no good scientist would ever try to base a claim upon a single report or observation, but would instead draw his conclusion from several independent observations.

This objection appeals to the principle of triangulation: certainty about the existence of a phenomenon is enhanced when the same object is viewed from different positions. The metaphor derives from navigation . In order to fix the location of a distant ship, it is necessary to take sitings from two different positions; the true position of the ship can then be located, at the intersection of the two lines of direction. By extension, a better picture of an object is said to result when you view it from (at least) two (different) positions. The gist of the triangulation objection is that only naive, unsupported claims to discovery rely upon a single viewpoint. Hence, the argument goes on, the deficiencies (horrors) said to characterize representation are merely deficiencies of bad scientific practice.

It is true that up to this point our argument has tended to concentrate on single representations, observations and knowledge claims, as if these occurred in isolation from other efforts and activities or, more generally, from the framework or context within which the claim takes place. We recall from the last chapter, however, that we stressed the importance of the social network for the constitution of a discovered object. In order to generalize our argument and thence demonstrate the holes in the triangulation principle, we need to broaden our consideration of the network of institutional arrangements and consider the constitution of objects within a *system* of discourse. We begin by examining the organization of discourse at the level of the text.

TEXTUAL ANALYSIS

In a famous article, Dorothy Smith examines the text of a report about the actions and behaviour of a (the reporter's) friend who was 'becoming mentally ill' [6]. She begins with the observation that the text can in fact be read in (at least) two quite different ways. This point dovetails the more general assertion about the vicissitudes of representation (in particular, the idea of indexicality) made in Chapter 2. Just as representations are underdetermined by objects in the world, so a text does not determine a particular interpretation (reading) of it.

The basis of this kind of textual analysis is that the practical expression of, or reference to, a phenomenon both recreates and establishes anew the existence of the phenomenon. In 'describing' a phenomenon, participants simultaneously render its out-there-ness. Smith makes explicit the congruence between practical textual expression (the way the text is organized) and the nature of the phenomenon (mental illness) which the text is 'about':

> The method of analysis assumes the structure of the conceptual scheme 'mental illness' which the reader uses in recognising 'mental illness' is isomorphic with that organising the text [7].

The implications of isomorphism between textual organization and textual phenomenon are extremely important. Firstly, there is no sense in which we can claim that the phenomenon (mental illness, in this case) has an existence independent of its means of expression. This means that it is nonsensical to attempt to arbitrate on the existence or otherwise of phenomena under analysis. Any attempt to understand discourse simply has to set aside this question. Secondly, the notion of isomorphism suggests, perhaps more clearly than we have yet been able to, not only that there is no object beyond discourse, but that the organization of discourse *is* the object. Facts and objects in the world are inescapably textual constructions.

These considerations lead Smith [8] to ask how the text can be read as a claim that the friend was becoming mentally ill. In this she concentrates on the effect of the organizational form of the text, rather than on the motives (interests, social circumstances) of readers and writers. What features of the organisation of the text make possible ('provide for') a particular reading? The work of Smith and of other analysts of texts [9] identifies four main features of textual organization which begin to answer this question:

(1) Preliminary instructions. Certain features of textual organization can be understood as initially providing the reader with instructions for making sense of the text with which s/he is faced. These features include the setting, headings, and textual openings.

The setting of an article, for example, its appearance in a 'serious' academic journal, can be read as an instruction that the text be taken seriously, that it be read as authoritative fact rather than fiction. Readers can orient to a series of vetting procedures which are presumed to have assessed and evaluated the text. The setting thus provides for the authority of the text by alluding to the (prior) involvement of others. The designation of the text (a 'Nobel Lecture' or 'research report') and mention of the institutional affiliations of its authors further suggest a prior network in which the claims of the piece have already been authenticated. The setting thus makes available the presence of other agents in the text: the reader may agree or disagree with what is said by the text, but its claims have been substantiated by at least some (significant) others before it reached the reader's eyes.

Headings, often in the form of titles, sub-titles and abstracts provide a category of attributes and actions which the reader can use to make sense of what follows. In other words, the instructions made available by the heading are: one ready way of making sense of the terms and items mentioned below is to find their relevance in terms of *this* heading. Headings also act as synecdoche: the text is to be understood as being about something abstractable from, and beyond, the text itself. This enhances the claimed antecedence of the objects 'upon which' the text reports. The significance of the text, 'in other words', is said to reside beyond the mere words in which it takes its form.

Headings also implicate the involvement of a legitimating agency prior to the reader. Not just in the obvious sense that writers have been at work prior to readers, but by offering interpretive categories for making sense of the lexical items in the text [10].

Textual openings can operate in a similar way to headings in that they suggest categories of relevance for a reader's interpretation of the text. For example, a tension between current and past knowledge is suggested, or between existing theories and new observations, and a solution suggested at the outset. This solution then represents the category of relevance in terms of which the reader is encouraged to make sense of the text. As Smith puts it, the main body of the text is offered as a reasonable puzzle for the solution which is presented at the outset [11]. If a mathematical proof was structured in this way we might complain that this was tantamount to assuming the answer. Instead, this feature of textual organization comes across as a reasonable way of presenting a solution ('fact'), which itself appears to pre-exist the text: 'Here is the solution, and here is how we came upon it'.

(2) Externalizing devices. Preliminary instructions encourage readers to orient to the involvement of others (agents) as authorizers of the facts (objects) being reported upon. As we noted when discussing modalizers, however, there is no simple equation between the enhancement of facticity and the presence or absence of agency. Preliminary instructions introduce the relevance of agents who can vouch for the legitimate claim being made, but who themselves are purportedly independent of the actual making of the claim. By contrast, descriptions of the making of the claim itself require rather different handling. The externalizing device provides for the reading that the phenomenon described has an existence by virtue of actions beyond the realm of human agency. The principal effect here is the non-involvement of human agency.

The discovered object is to be apprehended as neither the product nor the artful creation of scientists; scientists came upon these objects rather than creating them. But of course, the dilemma central to the whole tradition of scientific reporting is that scientists themselves are responsible for the reporting. The scientist needs to be the trusted teller of the tale but, at the same time, should not be seen as intruding upon the object.

One means of managing this dilemma is by using a textual voice which renders the actions of the scientist passive, and depicting entities like observations, results, information, lines of inquiry and so on as the prime movers. For example, 'the information then available suggested we were looking at a new kind of astrophysical source'; 'the results told me I was on the right track'; 'this line of inquiry brought us to the realization that. . .' [12]. In each case, human agents react passively to any or all of a class of inanimate entities: events, information, existing knowledge, facts, equipment, results and material circumstances. Although they may, especially in the case of great scientists, be particularly adept at seeing the implications of these events (results, circumstances, etc.), the scientists' contribution seems essentially coincidental with the unfolding realization about the objective state of the world. Any other scientist in the same situation would have been led to the same conclusion.

Externalizing devices include the invocation of community, a concept which is important in downgrading the contribution of any particular individual. The author is to be heard as having no special epistemological vantage point compared with the audience, for, once again it is emphasized that the character of the natural world is to be understood as indifferent to the number and variety of human agents engaged in its recording.

The notion of community is also commonly invoked by use of the 'royal we'. Thus, for example, 'we are all familiar with the twinkling of stars visible to the naked eye. . .'. This has the interesting effect of inviting the reader to become part of the existing state of knowledge. The constituency of the 'we' is not specified, yet the reader is encouraged to orient to an existing state of affairs, a state of existing knowledge which is shared by an unmentioned number of others. The invocation of these other witnesses to the knowledge (object) enhances the objectivity of this knowledge. It is not the idiosyncratic production of particular individuals, but a widely recognized (and hence objective) state of affairs.

The invocation of community membership legitimates the appearance in the story of other trusted tellers of the tale, usually by way of positive citations of their work. Their actions (findings, results) can provide a norm against which others' findings are to be evaluated, assessed and contrasted. There is often a cumulative effect of citing more and more people who similarily recognize the object being claimed. Once again the point is that the object is to be apprehended as objective in the sense that myriad members of a community recognized it as such; it is not an individual's idiosyncratic production.

(3) Pathing devices. The invocation of community is one important means of establishing the sanctioned recognition of existing past knowledge. More generally, a common feature of the organization of the scientific text, is the establishment of past states of affairs. Their pastness provides an important bedrock against which present interpretations can be made by the reader. In other words, establishing the fixity of a state of affairs (in virtue of its pastness) provides the reader with a framework for making sense of new observations and so on. Pathing devices are ways of fixing or establishing pastness, and of claiming the relevance of making an interpretation in terms of this pastness. They find expression most commonly in the depiction of links between existing (past) knowledge and the current state of affairs.

Preliminary instructions and externalizing devices provide for a reading of the out-there-ness of the object (discovery, finding) being related in a scientific text. The pathing device relates how this (purportedly) independent entity was brought within the confines of human agency, how it was 'captured'. (Hunting analogies are a popular metaphor for the belief that these objects pre-exist the efforts of humans to seek them out.) Part of the pathing device relies on the statement of the solution to the puzzle already provided in the preliminary instructions. Given the initial statement of the solution, the character of the discovered object, all the remainder of the text is to be heard, not just as a puzzle for the solution, but also as an inevitable route towards it.

A further effect of pathing is to ask the reader to suspend assessment of the individual textual facts and events until they have all been told. It is therefore inappropriate to raise a query about individual items, events or occurrences at the point of their appearance in the text. Instead we are encouraged to attend to the combined and cumulative effect of all the events. For example, if an account is being narrated it would be inappropriate to interrupt to ask a question about some mentioned event or action a few lines into the text. The narrator could well rebuff the interruption by pointing out that the import of this particular section would become clear later, that s/he was just giving the background at this stage and that this was not the main point of the story/report. (The same kind of effect is achieved by the insertion in a text of pointers to other parts of the text — 'see Chapter 5').

(4) Sequencing devices. Pathing devices organize the text into a collection of events and activities which are to be apprehended as having particular relevance for the asserted object. Sequencing devices are similar, but order the events in the narrative. Sequential organization thus acts as a 'cutting out' process, whereby other potential paths and other potentially relevant events and actions are backgrounded. The reader is encouraged to concur with the relevance of described events for the sequence in which they are located. Irrelevant events, in particular false leads, red herrings and blind alleys, are either excluded or are not tied to subsequent events. Sequencing devices thus provide for the connectedness of described events and activities.

ACTING AT A DISTANCE

Another way of approaching scientific discourse construes the work of explanation (description, reporting and so on) in terms of the relationship between the various kinds of statement produced by the scientist and the objects being explained. If an individual statement is merely linked to a single element of the world, little (or nothing) might be said to have been explained. If, by contrast, a single statement is linked with several different elements of the world, then we might begin to speak of an explanation taking place. This leads directly to the idea of power in explanations [13]. The more elements of the world that can be controlled by a single element of explanation, the more powerful this explanation is said to be. As Latour points out, the significance of this is in terms of the politics of explanation. Power here is analogous to political control; a single statement stands for (on behalf of) or represents many others at once. The explained elements are effectively subservient or neglible by comparison with the (elected?) representative. The point, however, is not that scientific discourse is motivated by a lust for political power. Merely that the power of explanation is an attempt to manage a practical problem. The aim is to act upon as many elements in the world at once, even though one is not physically in the same place as these elements. The practical problem, then, is how to act on these elements at a distance from them.

The idea of acting at a distance resonates with some of our earlier ideas about the relationships established by the text between observer, observation and observed object. For example, the rather arcane question about the sound of a distant falling tree (Chapter 4) can be understood as a problem of acting at a distance. How can a narrator relate what it sounded like when he is nowhere near the tree? The answer is that he needs to transplant the falling tree so as to bring it within his own narrative. In other words, he needs to re-align the relationships between narrator and object which are provided by the initial puzzle. This can be done, for example, by claiming that he had learned what it sounded like from some other person who had been there; or by producing a tape recording of the sound and so on. In all cases the attempt is to alter the original text by introducing another observer and by establishing a fresh set of relationships.

Clearly, a whole series of devices have the potential to act as additional observers: field notes, jottings, letters and other communiqués, other sociologists, their correspondence, photographs, video and sound recordings and so on. We would probably say the tape recording is the device which enables the narrator most persuasively to act on the setting while distant from it.

To successfully act at a distance is thus to possess the means to represent the distant element/setting without needing to be in the setting. But it is also clear that the observer needs to be able to claim that these representations do indeed emanate from the elements s/he is claiming. This generates three requirements of acting at a distance. Firstly, the observer needs to advance an account of the way in which the representation was made; he needs, in other words, to supply a story of the journey undertaken between the setting and the place from which he is now telling the story. In part, the means of relating the journey is supplied by the pathing devices mentioned earlier. So much the better if the journey is the rational, logically connected course of exploration. Secondly, he needs to convince his audience of the authenticity of the representation, that the representation, once made, has not been affected (deteriorated) in the course of the journey back to the place of the explanation. For this purpose, the claimant needs inscription devices which have the quality of what Latour has called 'immutable mobiles' [14]. For example, field notes must have the quality of not changing their meaning in the course of transportation between the field site and the academic audience for whom they are represented. Thirdly, the observer has to claim that his production of the representation is not intrusive upon the object itself; although he made the journey and fashioned the representation on site, he must make it clear that this did not change, alter or otherwise interfere with the pre-existing character of the object. Clearly, these requirements are to some extent in tension, for just the reasons we noted in our discussion of textual analysis. The scientist/observer has to be the trusted teller of the tale and yet his sole involvement in the representation must not be seen as impinging upon the object's character, i.e. his presentation is not just a distortion or partial reflection of what was actually the case.

CONCLUSION: (THE TRIANGULATION OBJECTION AGAIN)

The discourse of science is organised in such a way as to sustain and reinforce the objectivity of its objects, and systematically to diminish the contrary (constitutive) view. The device of splitting and inversion, and the use of modalizers, can significantly affect the facticity of a proposed claim (object) by insinuating the involvement of agency. By elaborating our discussion of modalizers we saw how whole texts are organized so as to structure associations and relationships between parties to the proposed object.

The work on textual analysis concentrates primarily on the textual management of the involvement of humans: How do texts deal with the fact that human agents are responsible for the instigation and construction of these reports and yet the objects they report are to be understood as beyond the hand of human intervention/construction? Preliminary instructions encourage readers to orient to the part played by unmentioned others in positively vetting the claims of the text's author(s); externalizing devices put the object beyond the reach (idiosyncratic fabrication) of human agency; appeals to community membership further authorize claims to facticity.

Analysis in terms of acting at a distance suggests we extend the concept of agency to encompass entities beyond just humans. We then see that the achievement of textual organization is to structure and align a whole series of relationships between objects, persons, equipment, resources, existing knowledge and so on [15]. In general terms, the text defines a 'moral universe' or network of relationships — which entities have which abilities, entitlements and obligations — within which the appearance of a new object (knowledge, fact) is shaped. In particular, acting at a distance implies an important difference between the character of the object being represented and that of the representation. Or, more significantly, the idea of distance between representation and object suggests these entities occupy discrete domains of discourse; consequently the representation and the represented object are prevented from contaminating each other.

As we shall discuss in Chapter 7, this last point has important implications for the pretensions of social science to adopt the discourse of the natural sciences. The discourse of the natural sciences tends to deny its objects a voice. Although electrons, particles and so on are credited with various attributes, they are constituted as incapable of giving opinions, developing their own theories and, in particular for our purposes, producing their own representations. The natural science discourse thus constitutes its objects as quintessentially docile and can act upon them at will. By contrast, various traditions in the social sciences wish to grant their objects a voice (and to refer to them as 'subjects'). This generates difficulties for the rhetorical constitution of distance. In particular, in the discourse associated with interpretivist social science, subjects/objects are granted the ability to talk back, have their own opinions and even to constitute their own representations.

We can now return to the triangulation objection. It was suggested that much of the argument for the constitutive position misleadingly implied that single acts of interpretation (representation) by an observer (claimant) were sufficient to constitute scientific knowledge. We have now elaborated the earlier discussion of single acts of interpretation to show that objects are constituted within networks of relationships. Representation (constitution) is never isolated from its textual context. The essence of the triangulation objection is that knowledge arises from *different* representations of the *same* thing. In other words, the principle of triangulation presupposes a constant extant object to which independent efforts of interpretation (measurement, description, etc.) are directed. However, it is apparent from our discussion, and in particular from the idea of acting at a distance, that sameness and difference are themselves accomplished in and through the discourse of representation. 'Sameness' or 'difference' is not an inherent property of (sets of) phenomena; it is instead a definition of a relationship between objects which is accomplished in virtue of their textual representation [16]. In other words, the triangulation objection is defeated since it is shown to rely on the presumption that objects can exist independently of discourse.

This chapter has introduced some of the various ways in which scientific discourse constitutes the character of the object it claims to be merely 'reporting'. The main conclusion is the constitutive point that the organization of discourse *is* the object. This conclusion seems at first sight to be very idealist. Surely, it might be said, it is absurd to argue that we cannot distinguish between a thing and what is said about that thing. But the constitutive position does not deny that participants (scientists and non-scientists) operate with distinctions between what they call merely textual descriptions and matters of fact; between what are variously thought of as 'words' on the one hand and the 'world' on the other. The point is that these distinctions are themselves part of the discourse. And these distinctions are applied in various ways. Frequently, one text (or set of texts) is said to comprise mere words whereas another text is deployed as a factual (real worldly) representations. This juxtaposition of distinctions is the basis of analytical irony. But given its overwhelmingly pervasive character, what resources can we draw upon to remind ourselves of the way in which textual organization persuades? We turn to this question in the next chapter.

NOTES

[1] D. Bloor, *Knowledge and Social Imagery,* Routledge & Kegan Paul, London (1976).

[2] B. Latour and S. Woolgar, *Laboratory Life: the construction of scientific facts*, 2nd edition, Princeton University Press, Princeton (1986).

[3] J. Law and P. Lodge, *Science for Social Scientists,* Macmillan, London (1984), p. 15.

[4] The difference between folk social science and social science is probably similiar to the difference between science and social science: one of resources. See Chapter 7.

[5] H. Garfinkel, M. Lynch and E. Livingston, 'The Work of a Discovering Science Construed with Materials from the Optically Discovered Pulsar, *Philosophy of the Social Sciences,* **11** (1981), pp. 131–158.

[6] D. Smith, 'K Is Mentally Ill: the Anatomy of a Factual Account', *Sociology,* **12** (1978), pp. 23–53.

[7] Ibid, p. 23.

[8] As if considerations had the power to lead anything or anybody! See page 75.

[9] For example, J. Gusfield, 'The Literary Rhetoric of Science',

American Sociological Review, **41** (1976), pp. 16–34; S. Woolgar, 'Discovery: Logic and Sequence in a Scientific Text', in K. D. Knorr, R. Krohn and R. Whitley (eds.) *The Social Process of Scientific Investigation. Sociology of the Sciences Yearbook,* **4** (1980), pp. 239–268.

[10] Cf. H. Sacks' concept of Member Categorisation Device (MCD). See H. Sacks, 'On the Analyzability of Stories by Children' in R. Turner (ed.), *Ethnomethodology,* Penguin, Harmondsworth (1974).

[11] Smith, op. cit. (note 6).

[12] Examples from Woolgar (1981), op. cit. (note 9).

[13] B. Latour, 'The Politics of Explanation: an Alternative' in S. Woolgar (ed.), *Knowledge and Reflexivity: new frontiers in the sociology of knowledge*, Sage, London (1988).

[14] B. Latour, 'Visualisation and Cognition: thinking with eyes and hands', *Knowledge and Society: studies in the sociology of culture past and present,* **6** (1986), pp. 1–40.

[15] The emergence of a recent body of work in the social study of science has developed this line of argument into what is referred to as 'actor–network' theory. For an analysis of scientific texts in these general terms, see the contributions to M. Callon, J. Law and A. Rip, *Mapping the Dynamics of Science and Technology*, Macmillan, London (1987).

A good exposition of actor–network theory is to be found in M. Callon, 'Society in the Making: the study of technology as a tool for sociological analysis', in W. E. Bijker, T. P. Hughes and T. Pinch (eds.), *The Social Construction of Technological Systems: new directions in the sociology and history of technology*, MIT Press, Cambridge, Mass. (1987), pp. 83–106; also J. Law, 'Technology and Heterogenous Engineering: the case of Portuguese expansion', in Bijker *et al*, ibid., pp. 111–134. See also B. Latour, *Science in Action: how to follow engineers through society,* Open University Press, Milton Keynes (1987); chapters by Callon and Law in Callon *et al.* (1987), op. cit.; J. Law (ed.), *Power, Action and Belief: a new sociology of knowledge*, Routledge & Kegan Paul, London (1986).

[16] Cf. M. J. Mulkay, *The Word and The World: explorations in the form of sociological analysis,* George Allen & Unwin, London (1985), Chapter 4; also H. M. Collins, *Changing Order: replication and induction in scientific practice,* Sage, London (1985), Chapter 2.

6

Keeping Inversion Alive: Ethnography and Reflexivity

Thus far, much of the argument has been devoted to advancing an inversion of the objectivist commitment associated with traditional conceptions of science: we have proposed that, rather than pre-existing our efforts to 'discover' them, the objects of the natural world are constituted in virtue of representation. Several other inversions have also been suggested: social norms provide an evaluative resource for the characterization of behaviour, rather than governing that behaviour; logic and reason are the consequence (often, the 'rationalization') of action rather than its cause; rules are resources for *post hoc* evaluation of practice, rather than the determinants of practice; facts are the upshot of knowledge practices, rather than their antecedents; and so on. Then, in the last chapter (5), we examined how scientific discourse functions to resist these inversions. The methodological horrors are resisted by systematically setting apart scientific objects from analytic practice and by establishing a 'moral order' which defines the rights and obligations of persons, objects and machines and which sanctions the relationships between these entities. Given its pervasive influence, what hope is there of challenging, let alone overcoming, the hegemony of this discourse?

In recent years, scientific discourse has come to be seen as the source of considerable obfuscation about the 'real nature' of science [1]. Much recent work in the social study of science starts with the notion that scientific discourse gives an essentially misleading or inaccurate picture of what actually goes on in science [2]. The first part of this chapter reviews the ethnographic study of science, an approach designed to counter misleading and idealized portrayals of science and scientific method by revealing the 'soft underbelly' of science: science as practised at the laboratory bench [3]. The second part of this chapter argues that the problems with this approach should lead us to develop a more reflexive perspective on the social study of science. We ask to what extent a reflexive approach might help sustain the kinds of inversion proposed in previous chapters. We ask, in other words, to what extent explorations in reflexivity might better enable us to subvert the moral order established by scientific discourse and the ideology of representation.

WHAT IS ETHNOGRAPHY?

In broad terms, ethnography is a style of research in which the observer adopts the stance of an anthropologist coming upon the phenomenon for the first time [4]. One takes the perspective of a stranger as a way of highlighting the taken-for-granted practices of the natives under study [5]. Ethno-graphy means literally description from the natives' point of view: rather than imposing one's own framework upon the situation, the ethno-grapher tries to develop an appreciation of the way the natives see things. In the case of science, our natives comprise the community of scientists. We adopt the perspective that the beliefs, assumptions and discourse of the scientific community must be perceived as strange.

Typically, the ethnographic study of science involves accepting a menial job in exchange for access to the research situation. In order to immerse oneself in the culture of the laboratory, the ethnographer offers help with such tasks as sweeping up, washing out test-tubes, preparing graphs and figures for publication and, perhaps, at advanced stages of co-operation and mutual trust, monitoring equipment during experimental runs. The general idea is to act as a spare pair of hands (and perhaps to provide a fresh pair of eyes) and thereby to become immersed in the culture. Over a period of eighteen months or so one becomes part of the day-to-day work of the laboratory.

The ethnographer takes notes, makes tape recordings (both video and audio), conducts interviews and amasses all the relevant documents that can be found in the laboratory. These latter would include, for example, draft calculations, paper jottings, traces, computer print-outs, memos between members of the laboratory, their correspondence with other scientists, the articles, books and reports which are being read and/or referred to and otherwise used. In the manner of an anthropologist doing fieldwork, the ethnographer of science collects the scriptures of the tribe.

The main rationale for this kind of work is that this process of collection and observation provides the basis for an authentic picture of what actually goes on in the laboratory. It is generally recognized that most existing accounts of what goes on in science are partial and distorted. Often these accounts have been culled from interviews with eminent ex-scientists or from other public pronouncements about the nature of science: they stress the methodical, systematic and logical basis of scientific procedure. By contrast, the *in situ* monitoring of scientific activity gives us the benefit of the experiences of an observer undergoing prolonged immersion in the culture being studied. This kind of participant observation thus makes it possible to retrieve some of the craft character of science. This approach is designed to reveal the messy, idiosyncratic, stop-and-start character of work in the laboratory.

The main burden of the ethnographic study of science is to take a 'stranger's' attitude to all aspects of the laboratory culture. To give an idea of the extent of strangeness necessary, and to indicate some of the difficulty in maintaining this stance, it should be stressed that this stance is applied to even apparently trivial objects. Contrast, for example, the following descriptions of a pipette, respectively from the point of view of native and ethnographer (stranger):

> A pipette is a glass tube with the aid of which a definite volume of liquid can be transferred. With the lower end in the liquid, one sucks the liquid up the tube until it reaches a particular level. Then, by closing the top end with finger or thumb to maintain the vacuum, the tube can be lifted and the measured volume of liquid within it held. Release of the vacuum enables the liquid to be deposited in another beaker, etc.

> Here and there around the laboratory we find glass receptacles, open at both ends, by means of which the scientists believe they can capture what they call a 'volume' of the class of substance known as a 'liquid'. Liquids are said to take up the shape of the vessel containing them and are thought to be only slightly compressible. The glass objects, called 'pipettes', are thought to retain the captured 'volume' and to make possible its movement from one part of the laboratory to another.

As we shall see below, this kind of relentless anthropological attention is necessary because it turns out that even such mundane objects encapsulate and sustain the culture of the laboratory, its beliefs, results and decisions from the past embodied in material artefacts [6].

Just as in any good anthropological investigation, the ethnographer of science must bracket her familiarity with the mundane objects of study and resist at all times the temptation to go native. The standard tension of any ethnographic study is present here. We want to see things from the natives' point of view but we don't want uncritically to adopt their belief system [7]. Note, however, that in one important sense it is more difficult to remain 'strange' in the exotic culture we call science than it is when conducting an ethnography of, say, the Navaho Indians. When the latter informants tell us that they are dancing in order to make it rain, we can readily draw upon scepticism which is 'in-built' in virtue of our membership of 'advanced Western culture'. But when informants amongst the tribe of scientists explain that the right-hand side of an equation 'follows' from the application of the rule of commutativity, we find it much more difficult to resist the apparent authority of this explanation. Why? Simply because respect for scientific rationality is deeply imbedded in our own (ethnographers') culture. In this sense and because, as we have suggested throughout, our own efforts at representation and study are themselves pale reflections of scientific practice, it is much harder to treat science as exotic. Rather like a Nahavo attempting an ethnography of the Navaho, the ethnographer of science has the problem of already being a quasi-believer.

In practical terms this problem amounts to learning when to ask the right questions and, perhaps more importantly, recognizing and noting the point when the answers are no longer surprising. For example, at an early point in an ethnography of solid state physics [8] I asked: Why do you affix the electrical contact to that particular point on the alloy sample? My informant adopted a condescending tone, clearly designed for the novice. As if merely reiterating 'what everyone knows' he explained: 'we need to minimize the resistance across the face of the sample'. Only a few days later I asked: 'How do you know that the temperature being recorded on the thermocouple is in fact what is indicated on the voltmeter over here?' Waiting with some trepidation, I was taken aback by the response: 'That's a good point. We should really check out whether or not that thermocouple is at the centre of the reaction or just on the edge of it'. Such episodes demonstrate the importance of ethnography as a learning experience. In making the transformation between novice and competent, one obtains a first-hand view of the structure of beliefs, taken-for-granted assumptions, what counts as a legitimate inquiry and so on.

RESULTS FROM ETHNOGRAPHIC STUDIES OF SCIENCE

The most striking feature of scientific practice is the extreme disorder of the laboratory: several observers report their surprise at the messiness of scientific work. The tidy, ordered image which percolates through idealized accounts of scientific procedure has little place in the cut and thrust of daily laboratory practice. In particular, it seems difficult to reconcile the moves made by scientists with the explicit application of abstract decision criteria (Chapter 1). In short, scientists' actions are highly indeterminate. Decisions about the kind of instrumentation to use, the types of experiment to run, the sorts of interpretation which are most appropriate, are all highly dependent upon local conditions, circumstances and opportunities. When rules of procedure *are* invoked, they tend to be used in a highly variable and often contradictory fashion.

The circumstantial press of affairs in the day-to-day life of the laboratory also means that decisions and activities are rarely undertaken in the manner of a dispassionate search for the truth. Scientists have little time for a reflective evaluation of the epistemological status of their actions and interpretations. 'Philosophizing' of this kind is most common among the elder statesmen of the field or among the disaffected and marginal members of the community. For the majority, the main and immediate aim is to make things work. Theirs is an instrumental rather than an epistemological concern. If you tell me that a certain alloy exhibits a glass transition temperature around the ambient temperature of the laboratory I might get excited, not because truth has been revealed, but because this piece of information enables me to set up a different kind of experiment, to apply for a research grant, or to once and for all defeat the claims of my competitors.

This instrumental orientation of scientists supports the conclusion, to use Knorr-Cetina's terms, that scientific activity is better termed constructive than descriptive [9]. In other words, scientists are not engaged in the passive description of pre-existing facts in the world, but are actively engaged in formulating or constructing the character of that world. This is fairly obvious from the reading and writing activities of our subjects: we see them constructing drafts, memos, letters and articles; they are also responsible for the production of computer print-outs, charts and line graphs [10]. It is perhaps less obvious that a whole series of decisions and assessments are imbued in the so-called 'raw' materials of the laboratory. The sample metals are chosen from among a variety of sources, the animals for testing are carefully selected and bred, the water used in experiments is purified, and so on.

Instruments and apparatus have a rhetorical neutrality in the sense that they are thought of as being merely 'used' or 'applied to' the materials (or animals) being investigated by the scientists. In many cases, these mechanisms are especially important because they appear to have the capacity to write nature 'automatically'. For example, there seems no obvious intervention between the alloy sample and the coefficient measurement displayed by a differential scanning calorimeter. But the selection and use of machines clearly involves scientists. More profoundly, many of the mechanisms are designed on principles established as the result of previous laboratory investigations. For example, the nuclear magnetic resonance spectrometer is not a neutral black box, but the embodiment of some twenty years of basic physics research. By merely 'using' the device, scientists invoke as neutral a mechanism which in fact draws upon, and is shaped by, a multitude of previous decisions, interventions and selections by previous communities of scientists.

The instrumentalism which guides scientists' actions in the laboratory, and the instrumentalism of the apparatus is closely tied to a yet more fundamental feature of scientific discourse. The very attributes of nature, the way in which the physical world is apprehended, described and classified, depend on the technologies which make these activities possible. As Knorr-Cetina puts it, scientific attributes are entirely constituted with regard to a possible system of instrumentation [11]. The 'temperature' of a substance is constituted by the availability of an expanding liquid metal (mercury) or by notions of heat flow across a potential difference (thermocouple). Previously solidified, and now taken for granted, decisions about these instruments define what temperature consists of. A 'discovery' that glass expanded and contracted instantaneously on contact with heat would revise what was to count as temperature [12].

Achievements and solidifications of the past thus provide the (purportedly neutral) technology (apparatus, instrumentation) whereby new attributes are themselves constituted. This theme can be recognized as a particular case of the more general argument that all observation is 'theory laden'. When (laboratory) scientists make observations, or produce results, they do so on the basis of a particular concatenation of past achievements. In Knorr-Cetina's terms, the scientific laboratory comprises materializations of earlier scientific selections; several orders of selections are embodied in each specific 'result'. The products which stabilize and solidify set the scene for the next set of interpretive decisions.

The rhetorical importance of materialization is that previous results are turned into a technology which, in the course of scientific practice, can be apprehended as a set of merely passive, neutral instruments. Facts are no longer just facts by virtue of their utterance; they are embodied in mechanisms which (are said to) enable further work (experiments, inferences, measurements, data collection). Facticity is thus enshrined in terms of instrumental value. As we shall see in the final chapter, it is along this dimension rather than in terms of any philosophical notion of the adequacy of procedure (method) that the natural and social sciences can be said to be distinguished.

Finally, the ethnography of science has shown that scientific acitivity is social in three main senses. First, it is clearly social rather than individual in the limited sense that much science involves teamwork; modern scientists are necessarily members of a team if not of a community of peers. And, of course, even individual action is necessarily oriented to a language community within which thoughts, actions and reflections take their meanings (see also Chapter 1). Second, scientific activity is social in the sense that all scientific actions are imbued with pre-selection. In particular, it is not possible, as was thought by earlier generations of sociologists, to distinguish the social from the technical (scientific) dimensions of scientific action. The esoteric scientific and technical details of scientific knowledge are themselves social. Indeed, one possible way of underlining this situation is to propose that the term 'social' simply be abandoned, lest it continue to be understood as denoting a special and discrete realm of phenomena [13]. This is, of course, a high risk strategy in that it presumes a similar (and simultaneous) abandonment of the special privilege attached to the notion of 'scientific'. Third, science is social in the sense that scientists direct their activities not towards 'nature' nor 'reality' but to the agonistic field: the sum total of the operations and arguments of other scientists. Nature and reality are the by-products rather than the predeterminants of scientific activity. This also enables us to see how science is infused with politics, not just in the limited sense of funding considerations or government or commercial interests, but in respect of an entire gamut of strategies of argument, mobilization of resources, rhetoric and so on. Negotiations as to, say, what counts as a proof in science are no more nor less disorderly than any argument between lawyers, politicians or social scientists.

PROBLEMS IN THE ETHNOGRAPHY OF SCIENCE

The results from the ethnography of science derive from a variety of case studies conducted in a wide range of different scientific research settings, ranging from plant protein research to high energy physics [14]. Not surprisingly, perhaps, many of these results concur with those of the recent social study of science more generally. However, a number of objections have been directed at the attempt to study science ethnographically. It is important to consider these because they highlight a basic weakness in the conception of ethnography employed to date.

One general objection is that focus upon the activities at the level of the laboratory bench is inappropriate because 'this is not where science happens'. The suggestion here is that the details of argument and persuasion are unsurprisingly no different from those of, say, car mechanics repairing an engine. It is only once these details are linked to science as an institution that it makes sense to say that we are focusing on science. Elsewhere this point finds expression in the warning to beware of ethnographic 'dazzle', the condition where the ethnographic observer loses sight of what all the observed details are details of. In another version, the complaint is that ethnography adopts the 'wrong unit of analysis'. It is asserted that sociologists should concentrate on the way in which relationships between groups affect the production of scientific knowledge, a question missed by concentrating on individual actions in a particular laboratory [15]. We thus see that although many sociologists seem to agree that the study of science 'as it happens' is a useful starting place, there is some disagreement as to what precisely can be gained from this [16]. As a consequence, a marked diversity of theoretical positions is urged upon ethnographers' observations of the strategies and manoeuvres of scientists at the laboratory bench. What accounts for this diversity? In order to broach this question we need to reconsider the basis for the attraction of studying science 'as it happens'.

In one sense of 'as it happens', the ethnography of science claims to produce a description of scientific work relatively unhindered by retrospective reconstruction. Contemporaneous monitoring of scientific activity enables the analyst to base her discussion on first-hand experiences rather than having to rely upon recollections made in the light of subsequent events. In a second sense, the study of science 'as it happens' enables the analyst to bypass intermediary constructions arising from reliance upon informants removed from their everyday working environments. *In situ* observation thus promises more direct access to events in the laboratory than, say, interview responses. In both cases, the idea is that more is to be gained from being on the spot than when attempting interpretation from a secondary perspective. It is claimed that *in situ* monitoring of scientific activity makes it possible to retrieve some of the craft character of science. In line with the call for a sociology which attends to the very content of scientific knowledge (Chapter 3), the ethnography of science promises close attention to technical details of scientific practice. The resulting picture of science portrays the scientist located firmly at the laboratory bench and treats with some scepticism the kind of representations provided by the scientist (and by spokespersons on behalf of science), especially where these are produced in situations removed (either temporally or contextually) from the scene of the scientific action.

The rhetorical appeal of the 'as it happens' clause is the promise of another means (method) of treating science as an object. And the ethnography of science claims not just to be different; it also promises a *more accurate* rendition of the object: a better, more persuasive way of 'acting at a distance' on science (cf. Chapter 5). This interpretation of the 'as it happens' clause thus reintroduces the basic assumption of the received view: ethnography can tell us what it's actually like in science since it provides a (new) way of overcoming obstacles to our apprehension of the phenomenon as it actually is. It is notable that the contrast between inaccurate or partial accounts of the nature of science and versions promised as the result of in-depth participant observation, often has considerable appeal for practising scientists (a fact which can be especially useful to the incipient ethnographer when negotiating access to the laboratory). In other words, the ethnographer's commitment to the 'as it happens' clause matches the tribe's own commitment to the idea of an actual (objective) state of affairs. The unearthing of the 'real' character of scientific practice is thought especially desirable as a way of countering 'distorted' or 'partial' accounts of 'what science is like'. In their eagerness to dispel 'deficient' versions of science, advocates of (this kind of) ethnography themselves subscribe to a discourse which underwrites objectivism.

The diversity of theoretical treatments associated with the ethnography of science reflects the force of the common assumption among sociologists of science about the importance of 'starting with the facts'. The implicit assumption of the neutrality of the ethnographers' observations means that they are regarded as 'facts for anyone', to be used for any of a variety of different purposes. The acceptance (usually *de facto*, rather than explicit) of the neutrality of sociological observations tends to have diverted attention away from the problem of what counts as the adequate use of representation when attempting to develop a critique of representation.

INSTRUMENTAL AND REFLEXIVE ETHNOGRAPHIES OF SCIENCE

We thus see that the variety of ethnographic study of science discussed so far (we can call it the 'instrumental' ethnography of science) can be charged with the same defect we have already identified in other varieties of sociology of scientific knowledge. On the positive side, these studies provide results and insights which contrast significantly with those provided by earlier sociologies of science; they provide alternative accounts of science which challenge our assumptions and basic preconceptions. However, these studies ultimately fail to challenge the core of the concept of science: they are conceived within, and fail to take issue with, the notion of representation.

Instrumental and reflexive ethnographies correspond to two quite separate conceptions of the nature and strategic significance of ethnography. Instrumental ethnography is most concerned with the production of news about science. In particular, this amounts to producing a version of science which, as we have seen, contrasts with extant versions (public or honorific accounts, published research reports and so on). In virtue of this kind of ethnography, it is suggested, science can be understood as essentially similar to non-science in most respects. The main conclusion is that science is an ordinary enterprise, neither to be feared nor accredited special (epistemological) status.

It is vital to recall, however, that an ethnographic report, like any report, is an exercise in persuasion. In other words, the persuasiveness of the ethnographer's account depends on the judicious use of a form of discourse which renders the object distant from (and hence antecedent to) the author of the report. The complication in the case of studying science is precisely that this involves the reduplication of the assumption which we have suggested is axiomatic to science, namely the assumption that representation can be deployed as a neutral means of apprehending pre-existing objectivities.

It follows that the more appropriate 'target' for ethnographic study is the practice of representation itself. In order to come to terms with the way in which representation pervades science, our approach should be reflexive since we need to explore ways of investigating our own use of representation. At the same time the notion of ethnography suggests we treat as strange the practices of representation as we engage in them. Hence, our ethnography should be a reflexive ethnography of representation rather than just an instrumental ethnography of science.

In a general way, this conception of reflexive ethnography resonates with attempts to develop interpretive ethnographies in anthropology [17]. In this latter tradition there is more to ethnography than just showing what, say, the Arawak are really like (by contrast, perhaps, with allegedly deficient versions of the same). Nor is the intention to add this piece of collected culture to a stockpile of stories about other 'primitive' peoples. This would be equivalent to generating a store of descriptions about different laboratories. Nor, again, is the point to use a picture of the Arawak to speak to theoretical concerns about the character of primitive peoples in general. Instead, the strategic value of reflexive ethnography is that it provides an occasion for reflecting upon, and perhaps reaching a greater understanding of, those aspects of our own culture which we tend to take for granted. As Geertz puts it:

The famous anthropological absorption with the (to us) exotic

... is ... essentially a device for displacing the dulling sense of
familiarity with which the mysteriousness of our own ability to
relate to one another is concealed from us. [18]

For Geertz, the critical target of ethnographic study is our own ability to
relate perceptively to one another. In terms of our current discussion, the
critical target is our own ability to construct objectivities through representation. These representational activities include the ability to adduce
evidence, make interpretations, decide relevance, attribute motives, categorize, explain, understand and so on. Stories about scientific practice, just
like stories about the Arawak, are most useful when they address these
fundamentals of representational practice.

The attempt to develop a reflexive ethnography raises some interesting
and difficult issues. A first problem is how to maintain awareness of one's
own reliance upon representation. Representation seems to induce a kind
of amnesia about itself: readers (and writers) are persuaded that they are
not being persuaded, that representation is a mere instrument for convey-
ing the world beyond. In the course of this argument itself, we could find
many places where opportunities for reflexive comment have been missed.
In the last chapter, for example, we were especially critical of the way in
which scientific discourse constitutes its object. Yet we proceeded to
articulate this criticism by treating 'scientific discourse' as an object for the
purposes of our own criticism. The many good existing analyses of representational practice (discourse) in science fail to attach any significance to
the fact that (traditional) sociological analysis itself depends upon the
unreflexive use of representation [19]. Or again, for example in Chapter 3,
we accused the strong programme of failing to see the implications of
scepticism about rule–following for its own attempt to specify guidelines for
the social study of scientific knowledge. Yet throughout the argument we
have ourselves advocated the application of two guidelines (inversion and
feedbacking) as a way of developing a radical challenge to representation.

A second, related, issue concerns the form which explorations of
reflexivity might take [20]. Although Geertz clearly identifies an important
target for reflexive inquiry (in our case, the fundamentals of representational practice), he is much less clear about what would count as an
adequate way of 'addressing' this target, let alone producing 'an adequate
understanding' of it. The idea of 'adequate understanding' suggests a fairly
conventional form of representation and persuasion. But this again fails to
attach significance to the observation that 'producing an understanding', in
the conventional sense, involves a relatively uncritical reliance upon
conventional forms of representation. By contrast, more recent work in

anthropology emphasises the textuality of ethnographic reports, that ethnographies are rhetorical performances whose authority is established through representational practice at the level of writing [21]. The approach to reflexivity suggested by this line of argument is an ethnography of the text [22].

A pessimistic response to this issue takes the line that since representation is all-pervasive, it is clearly not possible to try and evade it, so all attempts to 'step outside of' representation are doomed. But this response only follows if we conceive of explorations of reflexivity as attempting to escape representation, as seeking a way of side-stepping the ever-threatening methodological horrors (Chapter 2) and thereby more reliably grounding representational practice. Without wishing to try and escape representation, it is nonetheless worthwhile to pursue the possibility of developing alternative forms of literary expression. The idea is that this approach might modify existing conventions and thereby provide new ways of *interrogating* representation. The notion of interrogating representation contrasts with the aims of either *explaining* representation (as in those attempts to analyse scientific discourse which unreflexively redeploy the central axiom of representation for their own ends—see Chapter 5) or trying to *escape* it (which, we have already said, is impossible in the sense that all interpretive activities involve representation). Recently, several sociologists have begun to develop this kind of approach to reflexivity through the exploration of 'new literary forms' [23].

CONCLUSION

In this chapter we have looked at some recent attempts to develop an ethnography of science. Although useful as a way of questioning our acceptance of received views of the nature of science, it turns out that much ethnography of science has in important respects failed to get to grips with the most significant problem of science: the hegemony of scientific discourse. It was argued that this results from an instrumental conception of ethnography, a conception which reintroduces (or simply reaffirms) analytic assumptions which support the objectivist position.

We suggested instead the development of an alternative, reflexive perspective on science which self-consciously takes representation as its topic. It is this which may serve to keep alive the kinds of inversions suggested by the constitutive argument. However, at this early point in the exploration of reflexivity, the general question remains. To what extent can we develop a perspective which begins to provide adequate and effective resistance to the rhetoric of realism without slipping back into realistic rhetoric in the course of our own 'research'? In the final chapter we see how this question can be addressed in terms of the role of technology and agency in representation.

NOTES

[1] The locus classicus is P. Medawar, 'Is the Scientific Paper a Fraud?' *The Listener* (12 September 1963), pp. 377–378.

[2] This view is widespread and probably crucial to the critical moment claimed by social studies of science.

[3] B. Latour and S. Woolgar, *Laboratory Life: the construction of scientific facts,* 2nd edition, Princeton University Press, Princeton (1986).

[4] For example, M. Hammersley and P. Atkinson, *Ethnography: principles in practice,* Tavistock, London, 1983; D. Silverman, *Qualitative Methodology and Sociology,* Gower, Aldershot and Brookfield, Vermont (1985), Chapter 5.

[5] A. Schutz, 'The Stranger', *American Journal of Sociology,* **50** (1944), pp. 363–376, reprinted in Schutz, *Collected Papers II: Studies in Social Theory,* ed. A. Brodersen, Marinus Nijhoff, The Hague (1964), pp. 91–105.

[6] But how much anthropological attention is 'sufficient'? How strange is 'strange enough'? Although the latter description questions various key elements of the former, it is clear that the redescription opens up yet further opportunities for description. (What is meant by 'glass', 'receptacle', 'capture' and so on?) Here again we meet the 'horror' of inconcludability (Chapter 2); although the stranger stance provides a way of generating alternative descriptions, it cannot claim the superiority of these descriptions in terms of their exhaustiveness.

[7] On this particular point, the track record of recent ethnographers of science is poor: in possibly the worst manifestation of 'going native', two researchers have been married to members of the tribe they are studying.

[8] S. Woolgar, 'Time and Documents in Researcher Interaction: some ways of making out what is happening in experimental science', in M. Lynch and S. Woolgar (eds.) *Representational Practice in Science,* special issue of *Human Studies* (1988).

[9] K. Knorr-Cetina, *The Manufacture of Knowledge: an essay on the constructivist and contextual nature of science,* Pergamon, Oxford (1981).

[10] But see Chapter 7 on the rhetorical importance of mechanism.

[11] Knorr-Cetina, op. cit. (note 9).

[12] H. M. Collins, 'The Seven Sexes: a study in the sociology of a phenomenon, or the replication of experiments in physics', *Sociology*, 9 (1975), pp. 205–224.

[13] Latour and Woolgar, op. cit. (note 3), postscript to 2nd edition.

[14] For a list of these 'laboratory studies', see Latour and Woolgar, op. cit. (note 3), p. 285. For reviews of the field of 'laboratory studies' see, for example, K. D. Knorr-Cetina, 'The Ethnographic Study of Scientific Work: a constructivist interpretation of science' in K. D. Knorr-Cetina and M. Mulkay (eds.), *Science Observed: perspectives on the social study of science*, Sage, London (1983), pp. 116–140; S. Woolgar, 'Laboratory Studies: a comment on the state of the art', *Social Studies of Science*, 12 (1982), pp. 481–498.

[15] For a consideration of some of the main criticisms of laboratory studies, see Latour and Woolgar, op. cit. (note 3), postscript to 2nd edition.

[16] One example of such disagreement is the contrasting emphasis of Knorr-Cetina's work on theoretical issues in the sociology of knowledge and philosophy of science, and Latour and Woolgar's attention to the transformation of inscriptions. Knorr-Cetina, op. cit. (note 9); Latour and Woolgar, op. cit. (note 3).

[17] For example, C. Geertz, *The Interpretation of Cultures*, Basic Books, New York (1973).

[18] Ibid., p. 14.

[19] See M. Lynch and S. Woolgar, 'Sociological Orientations to Representational Practice in Science' in M. Lynch and S. Woolgar (eds.) *Representational Practice in Science*, special issue of *Human Studies* (1988).

[20] A wide variety of interests and practices are subsumed under the term 'reflexivity'. For an initial attempt to distinguish their main characteristics see S. Woolgar, 'Reflexivity is the Ethnographer of the Text' in S. Woolgar (ed.), *Knowledge and Reflexivity: new frontiers in the sociology of knowledge*, Sage, London (1988).

[21] J. Clifford and G. E. Marcus (eds.), *Writing culture: the poetics and politics of ethnography*, University of California Press, Berkeley (1986).

[22] Woolgar (ed.) (1988), op. cit. (note 20).

[23] For the most recent attempts to develop this work see M. Ashmore, *A Question of Reflexivity: wrighting the sociology of scientific knowledge*, University of Chicago Press, Chicago, forthcoming; M. Mulkay, *The Word and the World: explorations in the form of sociological analysis*, George Allen & Unwin, London (1985); and contributions to Woolgar (ed.) (1988), op. cit. (note 20).

7

Science and Social Science: Agency and Technology in Representation

We began by reviewing the wide variation in views about what counts as science and we noted the problems this caused for attempts to specify demarcation criteria (Chapter 1). Common to all these versions of science, however, is the idea of representation. Despite a series of severely damaging methodological and philosophical arguments about its impossibility (the methodological horrors), the idea of representation remains the crucial mainstay of objectivism (Chapter 2). The challenge to objectivism began when sociologists started to take seriously the relativity of scientific truths. The treatment of the generation of scientific knowledge as a sociological issue opened the black box; the sociology of science was transformed into a sociology of scientific knowledge (SSK) (Chapter 3). Attention to the nature of the connection between 'object' and its 'representation' was accompanied by critical analysis of the presumed direction of this connection: it was argued that facts and objects are constructed rather than discovered (Chapter 4). But we saw how the discourse of science is organized so as to resist this kind of inversion and to uphold the idea of representation (Chapter 5). Even (the less reflexive) ethnographic studies tend (unwittingly) to reaffirm the fundamental epistemological commitment to the idea of representation (Chapter 6).

Implicit in the argument throughout is the suggestion that the radical [1] implications of recent work in the sociology of scientific knowledge (SSK) are underdeveloped. Despite substantial achievements in the post-Kuhnian epoch, particularly impressive when judged against the legacy of earlier perspectives on science, this vigorous and increasingly influential body of research has yet to realize its full potential. SSK has drawn upon and elaborated major revisions in our basic preconceptions about science but has yet to explore the more radical consequences of this move. In particular, we have seen that the basic idea of representation remains essentially unscathed by most sociological investigations of scientific knowledge.

In the aftermath of SSK, we need to ask what comes next. The sociology of scientific knowledge has supplanted the sociology of science, but what will supplant the sociology of scientific knowledge [2]? Rather than just reiterating the arguments of the recent sociology of scientific knowledge — that science is a social process, that scientific method is not all it was once thought to be, and so on — we now need to consider what might lead us beyond the repeated application of the relativist–constructivist formula. At the end of the last chapter we suggested the need for an ethnography of what is taken for granted about representation: a reflexive exploration of our own practices of representation. This final chapter considers, in somewhat broader terms, the prospects for a vigorous challenge to the very idea of representation, and outlines some implications for social science.

ONTOLOGICAL GERRYMANDERING

The 'what comes next?' question is especially interesting because we take the view that relativism has not yet been pushed far enough. Proponents of relativism (both within and beyond SSK) are still wedded to an objectivist ontology, albeit one slightly displaced from its familiar venue. We have suggested, in effect, that the proponents of relativism are not dismantling representation *per se*, they are merely in the business of substituting sociological, literary and philosophical representations for the representations of science. Of course, this kind of substitution exercise is healthy enough as a beginning; if nothing else, it still raises the hackles of defenders of the supremacy of 'scientific method'. But it leaves the distinct impression that deeper, more fundamental questions remain unanswered. What accounts for the tangential character of the critique of science, the fact that the kernel of epistemological presumption remains essentially undisturbed, despite all the noise at the periphery?

A simple answer is that in at least two senses the relativist critique of science is itself scientific in its own aspirations. Firstly, the disciplinary origins of the social sciences lie in explicit attempts to mimic the aims, methods and achievements of natural science. The scientific pretensions of sociology owe much to the extent to which its founding fathers were impressed by the success of nineteenth-century biological sciences. Secondly, and more fundamentally, the disciplines which produce the critique of science share an important epistemological position. Although they might be characterized as distinct from science on disciplinary grounds (that is, in terms of their object), they share with science the *ideology of representation*, the set of beliefs and practices stemming from the notion that objects (meanings, motives, things) underlie or pre-exist the surface signs (documents, appearances) which give rise to them. A critique of science is empty or, at least, severely restricted, if it fails to address this ideology. The problem for science critics — if one regards it as a 'problem' — is that any attempt to dismantle this ideology, rather than a particular set of claims which emerge from a specific disciplinary (natural scientific) application of this ideology, appears tantamount to dismantling one's own discipline.

How then does the science critic deal with the spectre of apparent self-destruction? The relationship between science and those non-sciences which purport to provide a critical edge upon science is obviously crucial. It is fairly clear that the success (or, at least, the plausibility) of critiques of science is that they suppose (and present) themselves to be something separate (distant) from the science which is being deconstructed. This means that the course of deconstruction entails various implicit claims at differentiation between deconstructer and deconstructed. In short, the former presents the argument as if s/he was immune from the strictures applied to the target of the argument. This involves the subtle establishment and manipulation, in the course of the argument, of boundaries between those assumptions and arguments susceptible to deconstruction and those which are not. The relativist argument highlights the susceptibility to relativism of one set of claims and assumptions, while simultaneously backgrounding the fact that the very assumptions of relativism are themselves equally susceptible.

The practice of establishing and manipulating a distinction between arguments which are and are not susceptible to relativism has been called 'ontological gerrymandering' [3]. The importance of the role of agency is evident in this practice. The susceptibility to relativism of certain kinds of representation (labels, knowledge claims, definitions) is achieved by spotlighting the involvement of agency, and by emphasizing the possibility of arbitrariness, or distortion, of motivated, actively constructed representations. The critical gaze focuses on agencies such as the police and the courts in the case of the sociology of deviance [4]; on the news services in media studies [5]; and on the scientist in the sociology of science. Meanwhile, in all these studies, the role of the agent which makes these representations (about the police, etc.) is hidden, subdued, silent. In the terms of our discussion in Chapter 5, the insinuation and articulation of agency detracts from the facticity (representations) claimed by the subjects of study; at the same time, the deletion of agency enhances the facticity (representations) claimed by the author's own report. In the terms of Chapter 6, the strangeness, and hence constructed character, of the activities of the subjects are highlighted while attention to the activities of representer is played down, minimized and otherwise backgrounded. This difference between observer and subject/object is established and sustained throughout the course of texts which purport to be merely reporting upon the character of the other. Distance is effected by means of a rhetorical boundary between the constituting behaviour of others — to be regarded as strange and worthy of analysis — and the textual activities of the author — to be taken for granted as unworthy of attention. The essence of the claimed distance between the work of the text and the work of its subject/objects is that the text claims to operate at a separate level of discourse. The situation is analogous to the relationship between a photograph and its caption. The caption purports to be of a different order to the photographic image; we read the caption as directing our attention to actual characteristics of what is in the photograph, what it shows. The way in which the caption acquires its neutrality, its ability dispassionately to report upon a distant scene, is not an issue for the reader of the photograph.

Although the ideology of representation has been roundly criticized in general terms by a number of influential scholars (for example in philosophy of science [6]; in anthropology [7], as well as in social studies of scientific knowledge), it is important to note that most of these criticisms deploy a similar form of ontological gerrymandering: their authors produce texts which develop and elaborate arguments for the deficiency and/or historicity of conventions of representation; but the texts themselves trade upon the same conventions. Or, at least, they fail to address the extent to which this might be the case and whether or not it is consequential. Another way of making the same point is to observe that although the passive role of agency has been replaced by a more active conception of the scientist (police, news media, anthropologist), the 'author of criticism' remains a passive agent, depicted as disengaged and apparently immune from criticism.

THE IDEOLOGY OF REPRESENTATION AND THE ROLE OF AGENCY

Science is a highly institutionalized form of representational practice. But if the correct (and more challenging) target for deconstruction is the ideology of representation, rather than just 'science' as an organizational phenome-non, we need to remind ourselves that science is no more than an especially visible manifestation of the ideology of representation. Science is indeed, the tip of the iceberg of the modern obsession with technical rationality and reason; the public face of the ideology of representation. It provides, so to speak, the official party line on an attitude which pervades practices well beyond the confines of professional natural science. Discussions about science may be an obvious place to look for public pronouncements of this ideology, but science is by no means the exclusive locale for its operation. By insisting on a distinction between science (as object) and our own disciplines (as resource), we are in danger of mistaking relativistic critiques of science for an adequate appreciation of the more general phenomenon of representation.

As has already been hinted, the notion of agency is at the heart of the ideology of representation. The key relationship to be negotiated is that between the objects of the world and their representation through signs, records and so on. Agents of representation are those entities (actors, actants) which mediate between the world and its representation. Their role is presumed to be the relatively passive one of enabling or facilitating representation. However, there is an interesting asymmetry with respect to an agent's relationship to the world and to its representation. Agents are considered passive in the sense that they are not thought capable of affecting the character of the world. According to the ideology, the mediator does not intrude to the extent that s/he is any way responsible for the character of the de-sign-ated object. However, the agent *is* held responsible for the character of representations. While correct mediation amounts to author-itative (privileged) speech about the objective world, incorrect mediation can be said to the source of distorted representations of the (unchanging) world.

The alleged passivity of the agent *vis-á-vis* the facts of the world is captured in the idea that facts are neutral, that they are there to be discovered by anybody. But the alleged irrelevance of the agent provides an interesting awkwardness when it comes to acknowledging and rewarding individual scientists for their contributions to science. For these occasions provide a celebration both of the ideology of representation and the part played by the honoured individual. The dilemma is that the honoured individual is held to be especially capable of obtaining representations of the world, but that such representations do not arise solely in virtue of an individual agent. This accounts for the rather coy, 'lucky bystander' tone of Nobel Prize acceptance speeches: thank you very much, I couldn't have done it without the help of numerous others and *I just happened to be in the right place at the right time* [8]. The strength of the ideology of representation is the notion that, given the right circumstances, any other agent could equally have produced the same results, facts, insights and so on. This is the corollary of the view that the same facts were already there, enjoying a (timeless) pre-existence, merely awaiting the arrival of a transitory agent.

TECHNOLOGIES OF REPRESENTATION

However, as we saw in Chapter 5, the scientist is only one kind of agent held responsible for mediating between the world and its representations. The scientific laboratory is also populated by a wide variety of inanimate agents: experimental apparatus, oscilloscopes, measuring instruments, chart recorders and other inscription devices. Not all agents share equal responsibility in the business of furnishing representations of the world. Some are reckoned more capable than others, some particularly good at certain kinds of interpretive work, others as having outlived their usefulness, and so on. At any time, the culture of the laboratory comprises an ordered moral universe of rights and entitlements, obligations and capabilities differentially assigned to the various agents. This moral order can change with the introduction of a new agent into the community. For example, several person-days effort might be devoted to assessing the capabilities and performance of a newly purchased device for measuring changes in electrical resistance during isothermal annealing [9]. These deliberations might include negotiations between various agents (the company representatives, the head of the laboratory, putative users of the device) over the capabilities of the device. Even after its eventual purchase, the machine will be put through several further tests before gaining the trust necessary for its participation as an adequately socialized member of the community.

The hierarchy of rights and responsibilities includes the particular relationship between human and inanimate ˙agents of representation. Neophyte scientists also undergo socialization into the community, imbibing the ethos of representation, but learn to regard inanimate agents as 'machines', that is, as technologies of representation of a different order from them-Selves. While these machines are credited with the ability to produce direct ('automatic' or 'unbiased') representations of the world (Chapter 5), they are nonetheless presumed to remain under the control of the human agents.

The culture of scientific research thus comprises a moral order of entities we have called agents or technologies of representation. These entities are accredited with various capabilities on the basis of past results and solidifications of knowledge claims. An inscription device thus embodies the capability of making an apparently direct (or unmediated) connection between, say, the shape of the pulse on the oscilloscope and the character of the radiation received by a radio telescope. As we recall from Chapter 6, however, the significance of this is not just that prior results are taken for granted, but that the way the physical world is apprehended, described and classified depends on the technologies which make these activities possible. In other words, our knowledge of 'the way the world is' is shaped by the technologies of representation involved in our apparently neutral observation of the world.

REPRESENTATION 'OUTSIDE' SCIENCE

The pervasiveness of the ideology of representation is such that it informs the practices of critics of science as much as those of scientists. And this is even more the case to the extent that the critics of science argue that there is no essential difference between science and non-science. The critic of science is the mediator between the objects of his study (science, scientists' behaviour, etc.) and the signs (texts) which represent those objects, just as the scientist (observer) is the mediator between objects and their representation.

More generally, it can be seen that representational practices 'outside' science are also similarly informed by a moral order of representation. As writers, practical reasoners, conversationalists and so on, we do not conceive of our writings, reports and practical actions as merely whimsical products, having no connection whatever to 'a real world'. Despite the possible sources of distortion, bias and so on, we write (read, hear) with a commitment that signs are — at least potentially — the reflection of real entities in the world; that things other than the signs themselves lie behind and give rise to these (mere) signs [10]. The Cartesian dualism of representational ideology thus flourishes in practices outside of 'science' proper. Although it is often said that science can effect these connections more reliably, in virtue of the prolonged training of its (human) agents, the interpretive practices of everyday life nonetheless subscribe to the same ideology of representation. This is hardly surprising since science is the relatively recent manifestation of a philosophical tradition which began with the Greeks and which has increased its influence on all aspects of Western culture for over two thousand years.

Just as interpretive practices 'outside' science subscribe to the ideology of representation, so too are they dependent on the conceptions of agency already discussed. In particular, the available technologies of representation have the same important role in constituting the ('non-scientifically') represented world.

Walter Ong has argued that one particular class of technologies of representation, word-technology (writing, printing, electronic communication), have a profound effect on the nature of argument and reasoning [11]. What counts as an adequate argument is defined by the technology available for making that argument. By extension, what counts as an adequate representation depends upon the technology available for making that representation [12]. In other words, our apprehension of the way the world appears depends upon our acceptance of a particular technology as adequately 'representing' the world.

The point can be simply illustrated with an anecdote about the use of tape recording. I had planned to tape record some lectures to undergraduates as the basis for preparing materials for this volume. But I found that on two occasions the tape recorder had malfunctioned; the machine had generated only blank tapes. Although I was most immediately concerned about how to make up the loss, it became clear that there is a sense in which the 'content' of the lectures depend upon the means available for their recovery. The *absence* of a tape recording suggested (constituted) the existence of things which had gone on but which could not be recovered. If asked 'What was said in the lecture last week?', the lecture audience might have provided a whole series of different accounts, according to the resources available for its recovery: whether or not detailed notes were taken and so on. Each non-recorded lecture is now an (irretrievably) different object from the one we would have if there was a tape recording of it.

To generalize: what an occasion is (what the lecture consists of, what it is about, its content) changes according to the possible means of its representation ('recovery'). In other words, we *project* the existence of state of affairs (things that happened in the lecture) in virtue of an available technology. Obviously, this is quite contrary to our intuitive ideas about representation: the belief that objects in the world (what happened) pre-exist the devices we use to record them. In this case, the presence or absence of a recording device appears to have changed the character of the object being recorded. Just as we concluded in our discussion of laboratory practice (Chapter 6), attributes of a non-laboratory object are constituted in virtue of the technology available for its representation.

There are of course differences in the technologies of representation available to science and non-science. The extent and variety of inscription devices available to the social scientist and literary critic hardly compare with those used by science. Particle accelerators on the one hand; pens, typewriters and word processors on the other. But despite the difference in scale, the notion of agent and his/her relationship to the technologies of representation remains central. The main difference is one of resources, not of ideology.

CONFRONTING (INTERROGATING) REPRESENTATION

We have already suggested (Chapter 6) the exploration of reflexivity as one possible way of developing a thorough-going critique of science by attending to the concept of representation as we engage in its practice [13]. It now becomes clear that the exploration of reflexivity will have a particularly strategic importance if it focuses upon the moral order which sustains representation. The task is not just to understand the moral order which sustains the ideology of representation, but also to seek ways of changing it. We should note, in the spirit of Bloor's analysis (Chapter 3), that declarations of allegiance to an alternative ideology are unlikely to be sufficient. Instead, we need to look closely at the basis of our conventional practices of representation.

This raises the difficult question of what counts as effective criticism of the moral order within which one operates. Ong notes that critics of the latest form of word-technology (writing, printing, electronic communication) invariably find themselves in the position of having to adopt the form of technology they wish to criticize [14]. A critique of writing is most persuasive if it is written; attacks on printing are most effective if printed; the challenge to the dominance of the electronic media will be most efficient if conveyed electronically. The point of this intriguing observation is that critics are forced to adopt the form they wish to attack if they want their criticism to be taken notice of, because the 'latest technology' defines what is to count as persuasive, effective and efficient. By the time it is recognized and conceived as such, the 'latest technology' has already set the agenda for discussion about it. By this time it is already too late to try and propose equally worthy alternatives; for in becoming the 'latest technology', it has already carved out the niche in which it is uniquely qualified to operate. It is the best available alternative.

By analogy, the task for the next generation of 'social studies of science' is precisely that of seeking adequate and effective resistance in a situation where adequacy and effectiveness are defined by the ideology (representation) under critique. We recall from Chapter 5 that resistance reflects the judicious marshalling and alignment of resources; that the stuff of resistance (*res*) is precisely the cost of challenging and defeating each of the allies. It is clear, then, that the task is to see what can be done to reconstitute the moral order of representation, not only to explore alternatives to the current dominance of the rhetoric of realism, but also to dispute its right to define what counts as an alternative.

The ideology of representation is institutionalized in the sense that representational practices are normatively sanctioned. Practitioners (representers, interpreters) routinely display assumptions about the character of representation, and evaluate (both their own and others') representational practices by appealing to (often implicit) rules about correct procedure and method. In short, the ideology of representation provides a resource for the evaluation of interpretive practice [15].

Much is already known about fraud and deception in science [16]. Such instances undoubtedly constitute a violation of expectations about correct scientific behaviour. They do not, however, constitute a violation of the ideology *per se*. Fraud in science amounts to the distortion of findings and results, but it does not challenge the very distinction between results and the objects from which they purport to derive. For example, the 'bogus' claim to have discovered N-rays, while subsequently said to have been the upshot of improper behaviour, nonetheless trades upon the concept of separation between phenomena and their ostensive signs. It is thus useful to distinguish between deviance which violates consensual views about method and procedure, and deviance which violates the very idea of representation. While critics of science have been very good at pointing out the former, little has been done on the latter.

Garfinkel's prescription for revealing the fundamental structure of practical actions was to take ordinary, everday scenes and see what had to be done to cause trouble [17]. His notion was that these 'breaching experiments' expose the character of what was taken for granted. The degree of consternation caused by the breach is a reflection of the strength of adherence to the norms being violated. What then would constitute a breach of the ideology of representation? Clearly, we require some mechanism which suspends or makes problematic the distinction between object and its representation and, in particular, disrupts our assumptions about, or our inattention to, the role of the agent. This could be done, for example, by disrupting the supposed neutrality and authority of the agent, perhaps in the course of making this very argument! For example, the hidden hand of the author (observer) — the agent for presenting this argument — could be revealed at the point when readers least expect it [18].

CONCLUSION

We have seen that interpretive practices within and beyond professional 'science' are subject to an ideology of representation which sustains a

Cartesian dualism between objects and their representation. Subscription to the ideology of representation has produced a critique of science which is impressive by comparison with the treatment of science by earlier traditions, successful in providing alternative accounts of science, but ultimately unsatisfactory. By virtue of their inattention to the moral order of representation, most science critics to date have engaged in a form of ontological gerrymandering. The implicit acceptance of this moral order entails the uncritical use of various technologies of representation (notably, writing) which, as we have suggested, define the forms of argument accepted as persuasive. An alternative stance, which promises one way of building on the scepticism of the social study of science, is to seek ways of interrogating the moral order of representation in which we are currently locked.

Implications for social science
(1) How to stop worrying and live with SCIENCE.
One of the main implications of the work of the sociology of scientific knowledge is for attempts by researchers/scholars in 'non-scientific' disciplines to emulate the aims and achievements of the natural sciences. The debate about the SCIENTIFIC character of social sciences recurs from time to time — for example, in the German dispute about positivism — and is echoed in the tired old question found in almost all elementary introductions to sociology: Is sociology SCIENTIFIC? (the capitals denote the mythic, idealist connotations of this use of the term.)

In the context of initial introductions to the discipline, this question is often dropped as quickly as it is raised. The answer usually given hinges upon the extent to which sociology should emulate the natural sciences, especially in light of significant differences in the subject matter — sociologists study people rather than electrons! Even in more sophisticated treatments, the question is dealt with by means of comparing and contrasting the practice and aspirations of social and political sciences against a mythical model of SCIENCE.

Perhaps the most significant achievement of the social study of science is the finding that the natural sciences themselves only rarely live up to the ideals of SCIENCE [19]! The question about the extent to which sociology can or should emulate the natural sciences thus has a new twist. By recognizing the un-SCIENTIFIC character of both social and natural sciences, social scientists can stop worrying about how SCIENTIFIC they are. The question — 'Can social science be SCIENTIFIC?' — is misleading since science itself is not SCIENTIFIC except in so far as it represents itself as such.

(2) The apparent difference between science and social science is one of resources

Science and social science share the same ideology of representation. Public and other idealized accounts of science belie the overwhelming conclusion that representation in science is essentially no different from representation elsewhere. The discourses of science and social science are both organized to build the rhetorical distance between observer and observed object, and to establish the antecedence of the latter. On the other hand, it would be pointless to deny the perceived superiority of science. Its moral order seems stronger, the distance between observers and objects greater, better established. But the strength of a scientific explanation is no more than its degree of resistance to deconstruction. The difference between science and social science lies not in method but in the extent of resources invested to build and establish resistance.

(3) The need to search for alternative forms of social science explanation

The attempt by social science to emulate the achievements of natural science entails uncritical acceptance of the ideology of representation which, in turn, leads social science to seek to strengthen its explanations by maximizing the rhetorical distance between analyst and object. The problem, of course, is that social science also wants to grant its objects (which it calls its 'subjects') attributes and capabilities similar to those of the agent of representation (the analyst, author); the supposedly inherent difference between people and electrons is no more than a reflection of the attributes differentially granted these entities in the discourses of social science and science. This calls for a form of ontological gerrymandering which attempts to establish and reaffirm the exoticism (difference) of the subject/object. The activities and behaviour of the subjects of study are made strange and, in particular, are relativized (made to seem essentially arbitrary) by contrast with the normal, natural, taken-for-granted analytical activities of the observer/representer. As an alternative, we need to eschew forms of explanation committed to increasing the rhetorical distance between analyst and object. In other words, we need to play down the exoticism of the other.

(4) Self as a strategic target for social science

The perspective developed in this chapter calls into question one of the key claims made for the strategic value of relativistic studies of scientific knowledge. It is often argued that if scientific knowledge — widely held to be the hardest possible kind of knowledge — can be shown to be a cultural product, then all other (less hard) kinds of knowledge become relatively easy targets for social analysis. Or, to put it more exactly, if the general thesis of relativism holds in the substantive area where it is least likely to hold, this then establishes the general case for all other phenomena [20]. This is a corollary of the attempt to rescue scientific knowledge from its classification (at the hands of the classical sociology of knowledge) as a special case, exempt from consideration under the rubric of the sociology of (all other kinds of) knowledge. As we have seen (Chapter 3), the sociology of logic and mathematics is especially strategic in this regard, since these areas are often regarded as the hard core within scientific knowledge. However, our critical review of SSK shows that representation is even harder than scientific knowledge since Self (the analyst) has to assume a greater degree of hardness than scientific knowledge in order to effect its deconstruction. Sociological analyses of scientific knowledge need to constitute themselves (in the course of their argument) as harder than the scientific knowledge under study, even though the rationale for these analyses assumes the reverse. Analyses which rely upon conventional forms of rhetorical distancing seem actively to avoid the hardest possible case. The hardest possible case remains the Self: the disregarded agent of representation.

The call to find ways of interrogating Self is the fairly unsurprising conclusion to a critique of science with epistemologically radical pretensions. It is, after all, just the latest step in the long historical progression of de-centering - when displaced by Copernicus, Self found refuge in science; when displaced by relativism, Self once again found refuge in the analyst.

NOTES

[1] It should be clear that 'radical' here denotes an epistemological rather than a political radicalism. Much of the Marxist literature on science is disappointing in that its concentration on the distribution and resources of science appears to accept the fundamental auspices of science: the ideology of representation. See B. Latour and S. Woolgar, *Laboratory Life: the construction of facts*, 2nd edition, Princeton University Press, Princeton (1986), p. 277.

[2] This way of posing the question unproblematically adopts the rhetoric of progress. It is unclear to what extent practitioners in the sociology of scientific knowledge are willing to apply a constructivist view of progress to their own field of research. For an attempt to interrogate the notion of progress in this context, see S. Woolgar and M. Ashmore, 'The Next Step: an introduction to the reflexive project', in S. Woolgar (ed.), *Knowledge and Reflexivity: new frontiers in the sociology of knowledge*, Sage, London (1988).

[3] S. Woolgar and D. Pawluch, 'Ontological Gerrymandering: the

anatomy of social problems explanations', *Social Problems,* **32** (1985), pp. 214–227.

[4] For example, M. Spector and J. I. Kitsuse, *Constructing Social Problems*, Cummings, Menlo Park, California (1977); P. Conrad and J. W. Schneider, *Deviance and Medicalization: from badness to sickness*, Mosby, St. Louis (1980).

[5] For example, Glasgow Media Group, *Bad News*, Routledge & Kegan Paul, London (1977).

[6] For example, I. Hacking, *Representing and Intervening*, Cambridge University Press, Cambridge (1983); R. Rorty, *Philosophy and the Mirror of Nature*, Blackwell, Oxford (1980).

[7] See, for example, contributions to J. Clifford and G. E. Marcus (eds.), *Writing Culture: the poetics and politics of ethnography*, University of California Press, Berkeley (1986).

[8] For example, S. Woolgar, 'Discovery: logic and sequence in a scientific text', pp. 239–268 in K. D. Knorr, R. Krohn and R. Whitley (eds.), *The Social Process of Scientific Investigation*, D. Reidel, Dordrecht (1981); M. Mulkay, *The Word and The World: explorations in the form of sociological analysis*, George Allen & Unwin, London (1985), Chapter 8.

[9] S. Woolgar, *Science As Practical Reasoning*, forthcoming.

[10] In 'real world entities' I include considerations like 'what motivated the writer', 'what her background is' and so on; not just real world physical entities.

[11] W. Ong, *Orality and Literacy: the technologizing of the word*, Methuen, London (1982).

[12] This is an observation familiar to art historians. See, for example, S. Edgerton, *The Renaissance Discovery of Linear Perspective*, Harper & Row, New York (1976).

[13] See note 23 (Chapter 6).

[14] Ong, op. cit. (note 11).

[15] In line with our earlier argument (Chapter 3), it is important to note that neither rules nor ideology are here construed as determining interpretive practice.

[16] For a popular general account, see W. Broad and N. Wade, *Betrayers of the Truth: Fraud and Decceit in the Halls of Science,* Simon and Schuster, New York (1982). For an analytic treatment of the idea of fraud, see H. M. Collins and T. J. Pinch, *Frames of Meaning: the social construction of extraordinary science,* Routledge and Kegan Paul, London (1982).

[17] H. Garfinkel, *Studies in Ethnomethodology*, Prentice-Hall, Englewood Cliffs, New Jersey (1967).

[18] At this point the text could give way to a dialogue between author and his/her alter ego, thereby constituting a disruption of the univocal conventional textual form, a way of breaching expectations about normal modes of representation. However, since the whole point of a breach is its unexpectedness, and since the reader has already been forewarned of such an occurrence, the text does not follow this form here. Cf. S. Woolgar, 'The ideology of representation and the role of the agent', in H. Lawson and L. Appignanesi (eds.), *Dismantling Truth: science in post-modern times*, Wiedenfeld and Nicolson, London (1988).

[19] Indeed, as mentioned earlier, it is the argument of at least one philosopher (Feyerabend) that the natural sciences depend upon the active *transgression* of the rules of procedure and method to be found in conceptions of SCIENCE.

[20] H. M. Collins, 'Introduction: Stages in the Empirical Programme of Relativism', *Social Studies of Science,* **11** (1981), pp. 3–10.

Suggestions for Further Reading

CHAPTER 1: WHAT IS SCIENCE?

A useful overview of the different philosophical treatments of the demarcation question is given in A. F. Chalmers *What Is This Thing Called Science?*, Milton Keynes, Open University Press (2nd edition, 1982). See also S. Yearley, *Science and Sociological Practice*, Milton Keynes, Open University Press (1984).

A somewhat dated but useful collection of key writings in the classical sociology of knowledge is J. E. Curtis and J. W. Petras (eds.) *The Sociology of Knowledge: a reader*, Duckworth, London (1970). For a more analytic overview see P. Hamilton, *Knowledge and Social Structure: an introduction to the classical argument in the sociology of knowledge*, Routledge & Kegan Paul, London (1974).

Attempts specifically to relate the classical sociology of knowledge to the modern social study of science include M. J. Mulkay, *Science and the Sociology of Knowledge*, Allen & Unwin, London (1979); B. Barnes, *Interests and the Growth of Knowledge*, Routledge & Kegan Paul, London (1977).

CHAPTER 2: REPRESENTATION AND THE METHODOLOGICAL HORRORS

The main source for the sceptical interpretation advanced here is:

H. Garfinkel, *Studies in Ethnomethodology*, Prentice-Hall, Englewood Cliffs, New Jersey (1967).

However, this text is a notoriously difficult starting point. Most of the (introductory) secondary texts on ethnomethodology tend to underplay the sceptical reading advanced in this chapter. Among the better are:

D. Benson and J. Hughes, *The Perspective of Ethnomethodology*, Longman, London (1983); and W. Sharrock and R. J. Anderson, *The Ethnomethodologists*, Ellis Horwood, Chichester/Tavistock, London (1986).

The idea of taming the methodological horrors is initially developed in:

S. Woolgar, 'Irony in the Social Study of Science', in K. D. Knorr-Cetina and M. Mulkay (eds.), *Science Observed: perspectives on the social study of science*, Sage, London (1983), pp. 239–266;
and
S. Woolgar and D. Pawluch, 'Ontological Gerrymandering: the anatomy of social problems explanations', *Social Problems*, **32**, (1985), pp. 214–227.

CHAPTER 3: OPENING THE BLACK BOX: LOGIC, REASON AND RULES

The many critiques of the strong programme in the sociology of scientific knowledge include:

L. Laudan, 'The Psuedo-Science of Science?', *Philosophy of the Social Sciences,* **11** (1981), pp. 173-198.

S. P. Turner, 'Interpretive Charity, Durkheim and the Strong Programme in the Sociology of Science', *Philosophy of the Social Sciences*, **11** (1981), pp. 231–243.

More general attacks on relativism from philosophy are contained in:

M. Hollis and S. Lukes (eds.), *Rationality and Relativism,* Oxford, Blackwell (1982).

Some replies to these criticisms are located in two articles:

D. Bloor, 'The Strengths of the Strong Programme', *Philosophy of the Social Sciences*, **11** (1981), pp. 199–213.

B. Barnes and D. Bloor, 'Relativism, Rationalism and the Sociology of Knowledge', in M. Hollis and S. Lukes (eds.), *Rationality and Relativism*, Blackwell, Oxford (1982), pp. 21–47.

There are a large number of case studies of scientific knowledge which, in broad terms, are consistent with the strong programme. Two of the best are:

S. Shapin, 'The Politics of Observation: cerebral anatomy and social interests in the Edinburgh phrenology disputes' in R. Wallis (ed.) *On the Margins of Science: the Social Construction of Rejected Knowledge*, Sociological Review Monograph 27, Keele University (1979), pp. 139–178.
D. MacKenzie, 'Statistical Theory and Social Interests', *Social Studies of Science*, **8** (1978), pp. 35–83.

A collection of articles under the same rubric:

B. Barnes and S. Shapin (eds.), *Natural Order: Historical Studies of Scientific Culture*, Sage, London (1979).

For a critique of the explanatory form of case studies which follow the strong programme, see:

S. Woolgar, 'Interests and Explanation in the Social Study of Science', *Social Studies of Science*, **11** (1981), pp. 365-394; also ensuing correspondence in *Social Studies of Science*, **11** (1981), pp. 481–514.

A parallel tradition of case studies is based upon 'the empirical programme of relativism' and draws in particular on the work of Collins. See:

H. M. Collins, *Changing Order: replication and induction in scientific practice*, Sage, London, (1985).

A collection of case studies which follow the 'empirical programme of relativism' are in:

H. M. Collins (ed.) *Knowledge and Controversy: studies of modern natural science*, special issue of *Social Studies of Science*, **11** (1), (1981).

For some critiques of this particular tradition see articles by L. Laudan, K. D. Knorr-Cetina, D. E. Chubin and H. M. Collins in *Social Studies of Science*, **12** (1), (1982), pp. 131–143.

CHAPTER 4: INVERTING NATURE: DISCOVERY AND FACTS

On discovery:

A. Brannigan, *The Social Basis of Scientific Discoveries*, Cambridge University Press, Cambridge (1981), especially Chapters 1, 5 and 7.

T. S. Kuhn, *The Structure of Scientific Revolutions*, 2nd edition, University of Chicago Press, Chicago (1970), Chapter 6.

One of the major sources to predate current work in the social construction of scientific facts is:

L. Fleck, *The Genesis and Development of a Scientific Fact*, University of Chicago, Chicago (1979, originally 1935).

For the main case studies in the construction of scientific facts, see the references given under Chapter 6 (further reading).

CHAPTER 5: ARGUING SCIENCE: DISCOURSE AND EXPLANATION

There are several parallel traditions in the analysis of scientific discourse:

(1)

G. N. Gilbert and M. Mulkay, *Opening Pandora's Box: a sociological analysis of scientists' discourse*, Cambridge University Press, Cambridge (1984).

For a rationale for this approach, see also:

M. Mulkay, J. Potter and S. Yearley, 'Why an Analysis of Scientific Discourse is Needed', in K. D. Knorr-Cetina and M. Mulkay (eds.),

Science Observed: perspectives on the social study of science, Sage, London (1983), pp. 171–203.

A more recent and especially clear introduction to the relevance of this kind of 'discourse analysis' for a wider set of issues is:

J. Potter and M. Wetherell, *Discourse and Social Psychology: beyond attitudes and behaviour*, Sage, London (1987).

(2) Representation in scientific practice:
B. Latour and J. de Noblet (eds.), *Les 'Vues' de L'Esprit*, special issue of *Culture Technique*, **14** (1985).

M. Lynch and S. Woolgar (eds.) *Representational Practice in Science*, special issue of *Human Studies*, forthcoming, 1988.

(3) Analysis of texts using 'actor-network theory':
B. Latour, *Science in Action: how to follow engineers through society*, Open University Press, Milton Keynes (1987), especially Chapter 1.

The contributions to M. Callon, J. Law and A. Rip (eds.), *Mapping the Dynamics of Science and Technology*, Macmillan, London (1987), especially part II.

Also papers in J. Law (ed.), *Power, Action and Belief: a new sociology of knowledge,* Routledge & Kegan Paul, London (1986).

CHAPTER 6: KEEPING INVERSION ALIVE: ETHNOGRAPHY AND REFLEXIVITY
The main monographs in the ethnography of science are:

K. D. Knorr-Cetina, *The Manufacture of Knowledge: an essay on the constructivist and contextual nature of science*, Pergamon, Oxford (1981).

B. Latour and S. Woolgar, *Laboratory Life: the construction of scientific facts*, 2nd edition, Princeton University Press, Princeton (1986, originally 1979).

M. Lynch, *Art and Artifact in Laboratory Science: a study of shop work and shop talk in a research laboratory*, Routledge & Kegan Paul, London (1984).

For a listing of case studies following a similar approach see Latour and Woolgar (1986), op. cit., p. 285. See also note 14, Chapter 6.

Recent attempts to explore reflexivity and new literary forms include:

M. Ashmore, *A Question of Reflexivity: wrighting the sociology of scientific knowledge*, University of Chicago Press, Chicago, forthcoming.

M. Mulkay, *The Word and the World: explorations in the form of sociological analysis*, George Allen & Unwin, London (1985).

Contributions to S. Woolgar (ed.), *Knowledge and Reflexivity: new frontiers in the sociology of knowledge*, Sage, London (1988).

Index